# Satellite Remote Sensing
## *An Introduction*

**The Author**

Ray Harris has research experience of all the major satellite remote sensing systems, specializing in land applications in Britain and the Middle East. Until recently he was a lecturer in Geography at the University of Durham; he is now with the Environment and Space Systems Group of Software Sciences Limited in Farnborough, Hampshire. Dr Harris was for six years the Honorary General Secretary of the Remote Sensing Society, and won the Remote Sensing Society Eurosense Award in 1986.

GEOGRAPHY, ENVIRONMENT AND PLANNING

# Satellite Remote Sensing

## *An Introduction*

Ray Harris

Routledge & Kegan Paul
*London and New York*

*First published in 1987 by*
*Routledge & Kegan Paul Ltd*
*11 New Fetter Lane, London* EC4P 4EE

*Published in the USA by*
*Routledge & Kegan Paul Inc.*
*in association with Methuen Inc.*
*29 West 35th Street, New York,* NY *10001*

*Set in Sabon 10/12*
*by Hope Services, Abingdon, Oxon*
*and printed in Great Britain*
*by T J Press (Padstow) Ltd*
*Padstow, Cornwall*



*Library of Congress Cataloging in Publication Data*

*Harris, Ray, Dr.*
*Satellite remote sensing.*
*Bibliography: p.*
*Includes index.*
*1. Remote sensing. I. Title.*
*G70.4.H38 1987 621.36'78 87–4927*
*ISBN 0–7102–0305–5*
*ISBN 0–7102–1312–3 (pbk.)*

*British Library CIP Data also available*

*ISBN 0–7102–0305–5 (c)*
*0–7102–1312–3 (ppr)*

*To Mary*

# Contents

# Figures

# Tables

# Preface

Like John Lennon I was born at a very early age. I was brought up in Liverpool in the 1960s and from those days remember two themes of the times other than my Jesuit education: the Beatles and the stream of televised rocket launches from the United States. Commentators have said that we became bored with seeing the launches of the manned satellites. I never did. I remember with great clarity my emotional response to the climax 'Five.Four.Three.Two.One.Zero. We have lift off. We have lift off at 11.16 Eastern Standard Time'. I didn't even know what Eastern Standard Time was then, but enunciated with an authoritative American accent it sounded good. For the 1960s it all culminated in the Apollo 11 mission to the moon in 1969 and Neil Armstrong's small step for a man and giant leap for mankind.

Since then space has come a long way. Communications satellites have revolutionised not only our technical means of communicating with one another, but more importantly our concept of the world. Live cricket from Australia, golf from the United States, tennis from France, data transmissions by computers from Brussels to Detroit, bank transfers from Hong Kong to London, clear telephone calls to Oman, Tunisia, California and Rio de Janeiro: all these are daily events and taken for granted.

Remote sensing has benefited enormously from all this space activity. Satellite remote sensing is just one corner of space exploration and exploitation which encompasses commercial, industrial and scientific activities on a grand scale. This book has been written to provide an introduction to satellite remote sensing. It has been written with a reader in mind: the undergraduate who is taking a course in remote sensing as part of a degree in an environmental science discipline such as geography or geology. In fact during the writing I have often had individual students in mind who over the years have taken my 3H remote sensing option at Durham: they will recognise the material and the style, if there is one. My main apology to the reader is the date below: satellite remote sensing is an evolving field which is changing very rapidly indeed. Inevitably some of the book's contents will be out of date even by the time it is published.

*Ray Harris*
*Durham*
*September 1986*

# Acknowledgments

I owe a debt of gratitude to many people in the preparation of this book. In particular, Eileen Beattie and Kath Lund who typed the manuscript, and Arthur Corner, David Cowton, David Hume and Stephen Allen who drew the diagrams. I also want to thank Ian Simmons, Michael Blakemore, Derek Hudspeth, Sinclair Sutherland and my brother Brian for their help in different ways, and to acknowledge the stimulus and the intellectual environment to explore remote sensing provided by staff, colleagues and friends at Portsmouth Polytechnic, the University of Bristol and the University of Durham. The Middle East Centre of the University of Durham provided travel funds for research visits to the Middle East.

The author and publishers would like to thank the following for allowing imagery to be included in the text: Jet Propulsion Laboratory, California, USA (Figures 3.5 and 8.4), Natural Environment Research Council (Figure 4.6), European Space Agency (copyright Figure 5.2), University of Dundee (copyright Figures 5.3 and 5.4), Royal Aircraft Establishment (Figure 5.6), CNES (copyright Figure 5.9), MacDonald Dettwiler Associates, Canada (Figure 5.10), NASA (Figures 7.2 and 8.8), NOAA (Figure 10.2) and R.K. Raney (Figure 8.8).

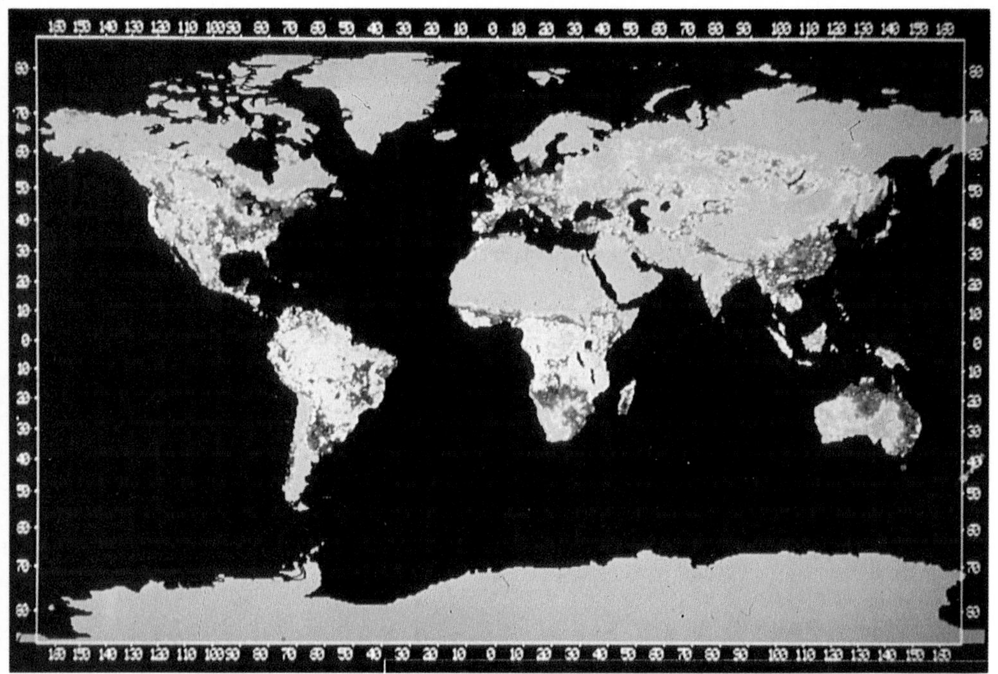

(a)

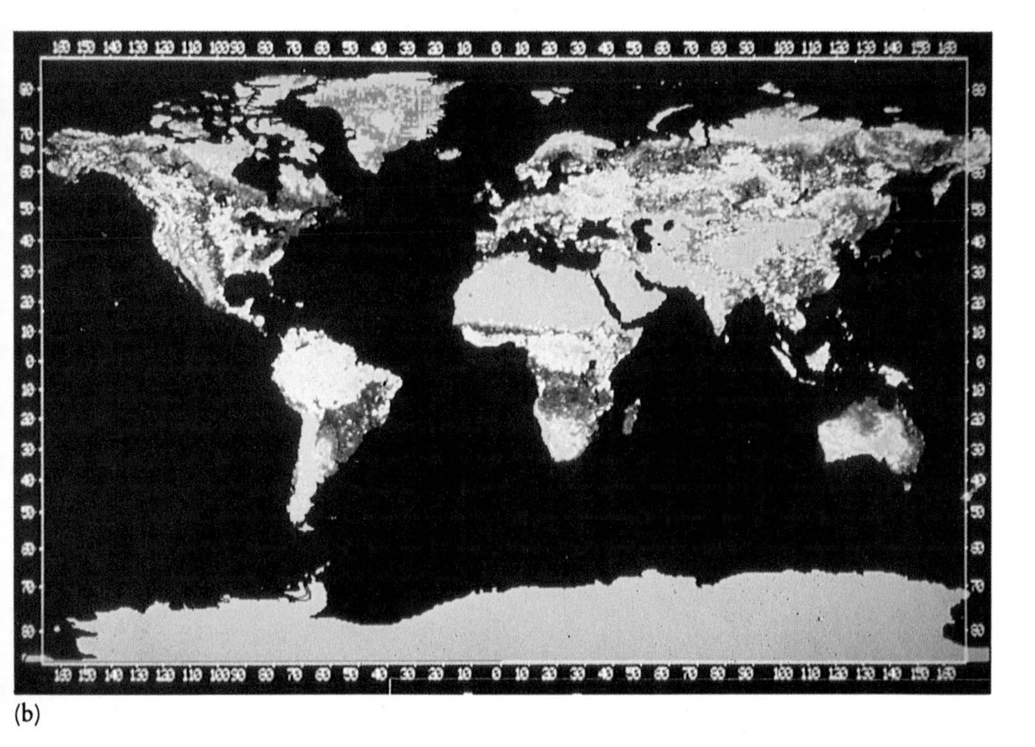

(b)

Global Vegetation Index (GVI) composite image for the Earth for four seasons:

(a) 12 April–15 May 1982

(b) 14 June–18 July 1982

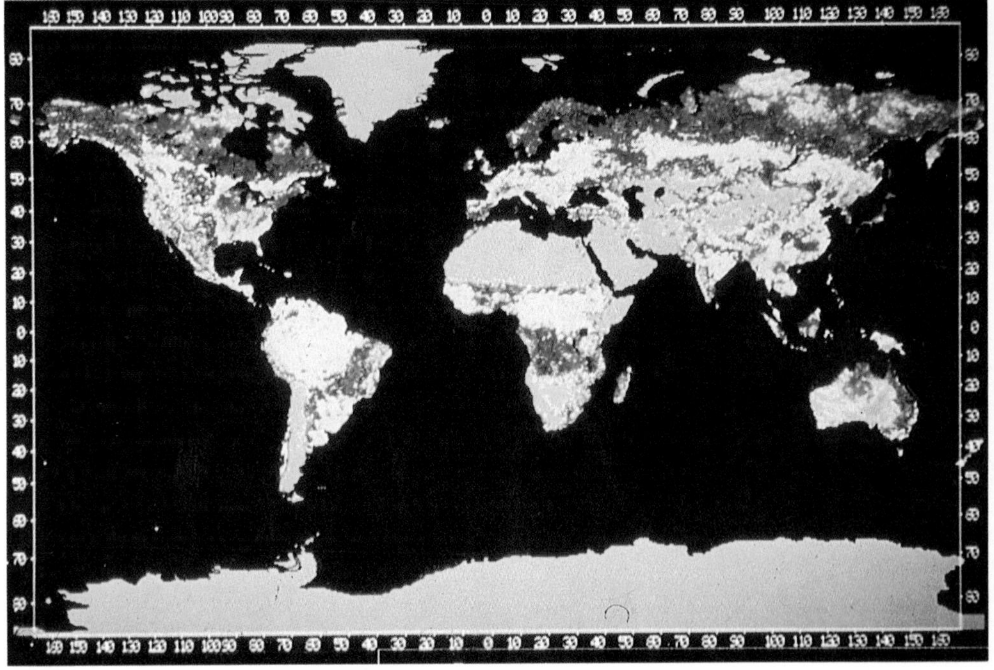

(c)

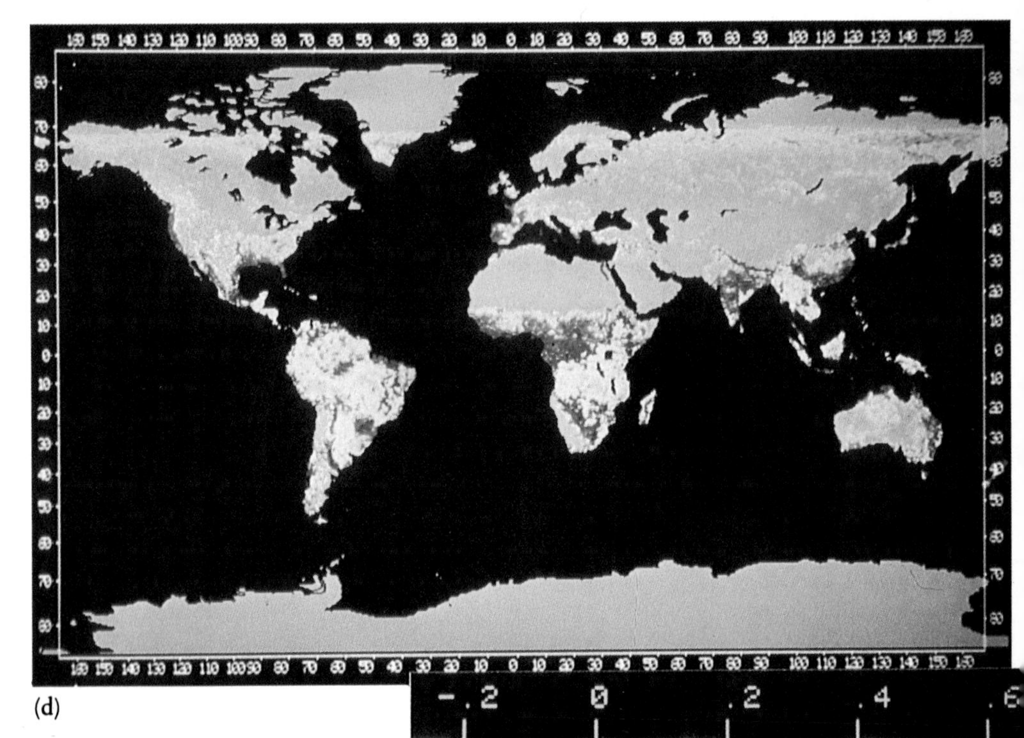


(d)

(c) 13 September–17 October 1982

(d) 13 December 1982–16 January 1983.

The range of colours and GVI values is shown in the lower part of (d): higher GVI values indicate greater vegetation activity. (From Justice *et al.* 1985)

*Chapter 1*

# Satellite remote sensing

The subject matter of satellite remote sensing is expanding at a very rapid and exciting pace. In only two and a half decades the exacting technology of Earth observation satellites has progressed from experimental and limited to quasi-operational and global. The next decade will see yet more operational satellite systems for Earth observation, with developments such as the polar platform and imaging radar systems. By the next century this little planet will be closely and continuously monitored by a band of satellites and sensors in space. This book is designed to give an introduction to what is happening in satellite remote sensing and how our knowledge of our environment is expanding at a colossal rate.

## Structure

The book is concerned specifically with remote sensing from satellite platforms orbiting in space. It is intended as an introduction to the subject for use by undergraduates studying remote sensing as a component of an environmental science subject, such as geography or geology. After this introductory chapter, the first part of the book (chapters 2 to 6) discusses the physical basis of the subject, the data employed and the ways in which they are processed. Chapter 6 on digital image processing links the first part of the book with the second part (chapters 7 to 10) which is concerned with the applications of satellite remote sensing in the environmental sciences. A concluding chapter looks at the way forward.

## Definitions and aims of satellite remote sensing

Satellite remote sensing developed from airborne remote sensing in the 1960s and 1970s. To define remote sensing is not a simple matter, and general definitions are often too broad. For example, Lintz and Simonett (1976) state that 'remote sensing is the acquisition of physical data of an object without touch or contact', and Barrett and Curtis (1982) suggest that remote sensing is 'the observation of a target by a device some distance away'. These two definitions are, like many others, too broad, and the first of the two could be construed to include reading a thermometer. Short (1982) gives a good if lengthy definition of remote sensing as

> the acquisition of data and derivative information about objects or materials (targets) located at the Earth's surface or in its atmosphere by using sensors mounted on platforms located at a distance from the targets to make measurements (usually multispectral) of interactions between the targets and electromagnetic radiation.

Satellite remote sensing is the use of sensors, normally operating at wavelengths from the visible (c. 0.4μm) to the microwave (c. 25cm), on board satellites to collect information about the Earth's atmosphere, oceans, land and ice surfaces. Commonly the information is collected in two-dimensional form either as a photographic image, such as the high-resolution images from the Metric Camera carried on the Space Shuttle, or as an array of digital data. In atmospheric and oceanographic applications the data collection may be one-dimensional, for example the vertical temperature profile of the atmosphere.

Satellite remote sensing may best be defined by its mode of operation. Remote sensing satellites orbit the Earth at a variety of altitudes from low polar (c. 200km) to high equatorial (c. 36,000km), and their sensors gather electromagnetic energy reflected, emitted or backscattered from part of the Earth–atmosphere system below the satellite.

## History of remote sensing

Remote sensing has a long pedigree. This section gives a brief review of the major phases in its development (Barrett and Curtis 1982). Five phases are identified, although they are a somewhat arbitrary division of an evolutionary process.

*Phase 1. Pre-1925* Some of the earliest photographs from the air were taken of Paris in the last century from cameras strapped to the breasts of pigeons (Curran 1985). Such photographs were uncontrolled as they were taken with a delayed shutter release after the pigeon had taken to the air. Aerial photography from balloons and in the twentieth century from aircraft showed the value of an aerial view of the ground. This value was highlighted in World War I when aerial photography was used to identify and plot enemy positions and military installations.

*Phase 2. 1925–45* The techniques of aerial photography saw widespread application in the inter-war period, particularly through topographic mapping from stereoscopic aerial photographs. As with many other aspects of geography, World War II gave a major impetus to the technology of aerial photography, and the geographical information thereby obtained, through improvements in the resolution of lenses and the stability of platforms. The Allies used aerial photography to great effect in identifying secret German rocket bases in the latter part of the war.

*Phase 3. 1945–60* Aerial photography became more widespread, and the applications of its products spread from topographic mapping to include applications in geology, agriculture, forestry and archaeology. This period saw aerial photographic coverage of many developing countries. For example, the Royal Air Force flew many photographic missions in the Middle East in the 1950s, and these photographs often still remain the best aerial-photographic coverage of large areas which is freely available.

*Phase 4. 1960–72* The birth of satellite remote sensing with at first relatively primitive cameras and sensors. The 1960s saw significant developments in weather satellites, and in the United States the weather satellite images improved greatly from the limited coverage by Tiros 1 in 1960 to an operational, global coverage by Essa and Noaa satellites at the end of the decade. Aerial photography had been characterised by being primarily concerned with the visible and near infrared parts of the electromagnetic spectrum. The satellite era saw investigation of other parts of the spectrum, notably the thermal infrared.

*Phase 5. 1972–Present* This period has seen a maturity in satellite remote sensing, a development of high spatial resolution sensors for land applications, operational data collection of atmospheric information, and the experimentation with new sensors in a variety of wavebands. The period began with the launch of the first Landsat satellite in 1972, and in the mid-1980s the Landsat programme has become semi-operational and been transferred into private ownership, and the first operational French SPOT satellite has been launched.

The history of satellite remote sensing has been one of increased data capture in terms of space, time and radiance. That is, larger areas covered with higher spatial resolution more frequently, and using more of the electromagnetic spectrum for data collection. But why? Why has this data explosion taken place in satellite remote sensing, and why does satellite remote sensing provide a useful source of environmental information? Barrett (1974) listed some of the reasons why satellites are a useful source of climatological information. His list can be adapted to apply more broadly to satellite remote sensing of the planet Earth.

(1) *Data coverage* is greatly improved. Satellite remote sensing allows data collection in remote land, ocean and ice areas which otherwise might be data voids or have a limited data supply for reasons of accessibility or for political reasons.
(2) *Homogeneity* of data. Information collected by any one satellite system (for example, the Landsat Thematic Mapper) uses one sensor and so provides spatially consistent data, while ground-data collection uses a variety of methods of data capture. For example, land-use surveys in adjacent countries nearly always differ, so matching of land-use maps at political boundaries leads to difficulties of interpretation.
(3) Satellite remote sensing data are *spatially continuous* compared to the point or small-area sample data commonly found in (say) meteorology, oceanography and geology. Satellite remote sensing provides the best quality large-area coverage data base on the Earth yet in existence.
(4) The data from satellite remote sensing are commonly in a form suitable for *computer processing*. Sensors other than spaceborne film cameras produce digital data which are in a form suitable for direct computer processing. In fact the vast quantities of data produced by satellite remote sensing demand rapid computer processing. For example, a Landsat Thematic Mapper scene has 6500 rows of pixels, 6920 pixels per row and seven wavebands: a total of 273 Mbytes of data for just one scene.
(5) The *frequency* of data collection is greatly improved using satellite remote sensing. Geostationary weather satellites scan a disc of the Earth every thirty minutes, polar orbiting weather satellites collect data twice per day for each part of the Earth, Landsat satellites pass over the same point on the Earth every sixteen days, and SPOT can view the same part of the Earth on successive days by off-nadir viewing.
(6) The *time base* of a single pass of a satellite is very restricted, so that spatial changes in environmental variables are minimised. For example,

the Landsat Thematic Mapper images a scene of 170km × 185km in 25.87 seconds.

(7) Measurements from satellites are *complementary* to conventional observations. *In situ* observations are needed to calibrate satellite observations, and satellite remote sensing allows a spatial extension of ground measurements which would be impossible without extensive and costly ground surveys (Gregory 1985).

(8) Satellite remote sensing provides a *low cost* means of environmental data collection, although in the 1980s the costs of Earth resources satellite data have increased steeply. The computer compatibility of satellite remote sensing data allows reduction in the time and manpower required to produce environmental survey information.

The emphasis in this section has been on data acquisition, and therein lies a limitation of satellite remote sensing. The definition given by Short (1982) earlier states that remote sensing is 'the acquisition of data and *derivative information* about objects or materials (targets) located at the Earth's surface or in its atmosphere' (my italics). Satellite sensors provide radiance data, but this alone does not provide what most users require. Environmental applications of satellite remote sensing require the processing of radiance data into derivative products such as maps of land use, geology or water and atmospheric pollution so that the extensive body of radiance data collected by satellites can be used to achieve their potential.

## Sources of information on satellite remote sensing

Satellite remote sensing is replete with abbreviations and acronyms. A list of some of these is given in Appendix 1 to provide a source for reference and to guide the reader through this alphabet soup.

A wide range of information on satellite remote sensing is published in formal and informal ways. Excellent collections of Landsat images have been produced by Charles Sheffield as *Earthwatch* (Sheffield 1981) and *Man on Earth* (Sheffield 1983). A satellite image atlas of the UK has been prepared by Bullard and Dixon-Gough (1985), and Fotheringham (1979) has produced a collection of weather satellite images.

Much scientific remote sensing material is published as conference proceedings, but three scientific journals extensively publish papers on satellite remote sensing: *International Journal of Remote Sensing*, *Photogrammetric Engineering and Remote Sensing*, and *Remote Sensing of Environment*. In addition, two parts of the *IEEE Transactions* contain papers on satellite remote sensing, largely of a technical nature: *Geoscience and Remote Sensing* and *Systems, Man and Cybernetics*. Other regular publications of a broader nature which include relevant material on satellite remote sensing are *Flight International*, *Space* and *Space Markets*.

A number of newsletters are published which contain relevant and up to date information on the subject. Some of the newsletters are detailed below.

*Earth Observation Quarterly* published by the European Space Agency, ESTEC, PB 299, 2200 AG Noordwijk, The Netherlands.
*EOSAT Newsletter* published by EOSAT, 4300 Forbes Blvd, Lanham, MD 20706, USA.
*Landsat Data Users Notes*, the predecessor to the *EOSAT Newsletter* but discontinued in March 1986. Available on microfiche from Landsat Customer Services, Mundt Federal Building, Sioux Falls, SD 57198, USA.
*National Remote Sensing Centre Newsletter* published by the NRSC, Space Dept, Royal Aircraft Establishment, Farnborough, Hampshire, GU14 6TD, UK.
*Remote Sensing News and Letters* published by the Remote Sensing Society, Department of Geography, University of Nottingham, Nottingham NG7, UK.
*SPOT Newsletter* published by SPOT Image, 16 bis avenue Edouard Belin, F31055, Toulouse Cedex, France.
*Washington Remote Sensing Letter* published by Dr Murray Felsher, PO Box 2075, Washington DC 20013, USA.

In addition to these, *Geographical Abstracts Part G*, published six times per year by Geo Abstracts Ltd, 34 Duke St, Norwich NR3 3AP, UK, contains c. 2500 citations and abstracts annually concerned with remote sensing, photogrammetry and cartography.

## Satellite remote sensing

Satellite remote sensing has seen a very broad spectrum of applications. Weather forecasting and disaster warning using meteorological satellite information are now routine. In the 1970s the USDA and NOAA used meteorological and Landsat data in a joint programme entitled the Large Area Crop Inventory Experiment (LACIE) and its successor Agristars (see chapter seven), which allowed prediction of grain crop production in North America, the USSR and several other countries. In 1982 at UNISPACE in Vienna NASA announced that one theme of its future activities would be

Global Habitability, which would ask which components of the Earth's biogeochemical cycles could be analysed using satellite remote sensing in conjunction with other data sources. This theme has been echoed in the International Satellite Land Surface Climatology Project which aims to assess changes in climate consequent upon changes in land surface properties, some of which may be caused by man. NOAA regularly produces composite vegetation index images of the globe from the Advanced Very High Resolution Radiometer data received from its weather satellites. Satellite remote sensing data are regularly used for ocean-ship routing and for producing sea-ice charts.

These activities illustrate the progress of the technology towards a complete Earth observing system. Meteorological satellites have provided information operationally since the 1960s, and in the 1980s we are seeing the growing maturity of satellite remote sensing for land and ocean applications. The technology will continue to develop and mature together with the applications of the information. This book provides an introduction to the cuisine of satellite remote sensing and gives a flavour of the range of applications.

*Part I*

# Energy, Sensors and Satellites

# *Chapter 2*

# Energy characteristics

This chapter discusses some of the characteristics of electromagnetic energy which are relevant to satellite remote sensing. The chapter is divided into four sections. The first two sections discuss plane waves and blackbody radiation, and introduce the nature of the energy which satellite remote sensing detectors employ. The second two sections discuss the absorption of energy by the atmosphere and the reflectance of that energy by the surface upon which it falls: both of these topics are fundamental to the subject of this book. For those not mathematically inclined do not be put off by the equations in the blackbody radiation section. For an exhaustive discussion of this subject matter see the *Manual of Remote Sensing* (Colwell 1983).

## Plane waves

Electromagnetic (EM) radiation travels in the form of waves, and these EM waves reflected by or emitted or backscattered from objects give us information about the objects themselves. Plane waves are the forms of EM wave energy in free space and are characterised as having a constant phase over a plane perpendicular to the direction in which the wave is travelling. Plane waves have a wavelength ($\lambda$), an amplitude (A) and a phase (L), and these characteristics are shown in Figure 2.1. Plane waves all travel at the velocity of light $c$, which is approximately 300 million metres per second ($ms^{-1}$), and have wavelengths which range from $3 \times 10^{-9}$m (gamma rays) through visible light ($0.4–0.7 \times 10^{-6}$m), to very long radio waves of $3 \times 10^{6}$m wavelength. As their velocity is constant at the speed of light then the frequency of plane waves is inversely proportional to their wavelength. That is, as the wavelength becomes shorter then more waves pass a point in

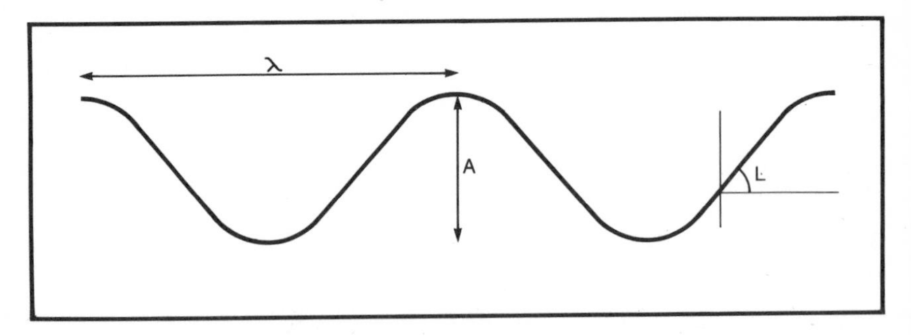


2.1 Plane waves having wavelength λ, amplitude A and phase L

a given time period so the frequency is greater. Visible and infrared remote sensing normally use wavelength to characterise the range of plane waves, but work in microwave remote sensing commonly employs both wavelength and frequency. Table 2.1 gives the wavelengths and, where appropriate, frequencies commonly found in remote sensing. Table 2.1 uses both micrometer (μm) and nanometer (nm) as both are used to describe the wavelengths of visible and infrared EM radiation:

$$1\mu m = 10^{-6}m \text{ (one millionth of a metre)}$$
$$1nm = 10^{-9}m \text{ (one billionth of a metre)}$$

*TABLE 2.1: Wavelengths and frequencies used in satellite remote sensing*

| *Type* | *Wavelength* | | *Frequency* | *Example* |
|---|---|---|---|---|
| Visible | 0.4– 0.7μm | 400– 700nm | — | SPOT HRV |
| Near infrared | 0.7– 1.5μm | 700– 1500nm | — | Noaa AVHRR |
| Middle infrared | 1.5– 3.0μm | 1500– 3000nm | — | Landsat TM |
| Thermal infrared | 8.5–12.5μm | 8500–12500nm | — | Meteosat |
| Microwave | 10–300mm | | 1–12.5GHz | |
| X-band | 24–38mm | | 8–12.5GHz | |
| C-band | 38–75mm | | 4–8GHz | ERS-1 |
| L-band | 150–300mm | | 1–2GHz | Seasat |

## Blackbody radiation

All bodies with temperatures above zero degrees Kelvin (0K = −273 °C) emit EM radiation. The temperature of the object determines the wavelength of maximum EM energy emission by Wien's displacement law

$$\lambda_{max} = \frac{a}{T} \tag{2.1}$$

where $\lambda_{max}$ is the wavelength of maximum energy emission, T is the temperature of the object (deg K), and a is a constant (2898μm K). By using Wien's law the temperature of a remote object can be measured by observing its spectrum and identifying the wavelength of maximum energy emission ($\lambda_{max}$). The hotter the body then the shorter the wavelength of maximum energy emission. For example, the $\lambda_{max}$ of the Earth with a surface temperature of 288K is approximately 10μm, while the $\lambda_{max}$ of the sun with a surface temperature of c. 6000K is approximately 0.55μm. This relationship is used in thermal infrared sensors on board satellites to measure the temperature of cloud tops, land and ocean surfaces, and can be particularly useful in geology in identifying geothermal areas.

A blackbody is not necessarily black but an object whose emissivity ($\epsilon$) is 1; that is, it emits all the energy it absorbs. The Earth and the sun are blackbodies. For a blackbody the relationship between emission of EM radiation and its frequency is defined by Planck's radiation law

$$E = hf \tag{2.2}$$

where E is the radiant energy, f is the frequency of radiation, and h is Planck's constant. Planck employed experimental data to determine the energy radiated by a blackbody as

$$E_\lambda = \frac{2\pi hc^2}{\lambda^5} \left( \frac{1}{e^{hc/\lambda kT} - 1} \right) \tag{2.3}$$

in $Wm^{-2}$ $Å^{-1}$, where

h = Planck's constant ($6.266 \times 10^{-34}$Js)
k = Boltzmann's constant ($1.38 \times 10^{-23} JK^{-1}$)
c = velocity of light ($3 \times 10^{8} ms^{-1}$)
λ = wavelength of energy (Å)
T = absolute temperature of the blackbody (deg K)

Figure 2.2 illustrates this relationship in diagrammatic form, and it is clear from this that hotter bodies emit more energy at a shorter wavelength than do cooler bodies.

Equation 2.3 is written for one wavelength λ. It can be rewritten for all wavelengths to give the total power radiated per unit area of a blackbody

$$E = \frac{2\pi^5 k^4}{15c^2h^3} \; T^4 \tag{2.4}$$

in $Wm^{-2}$. The first component of the right hand side of the equation consists entirely of constants (π, k, c, h), so the equation can be simplified to

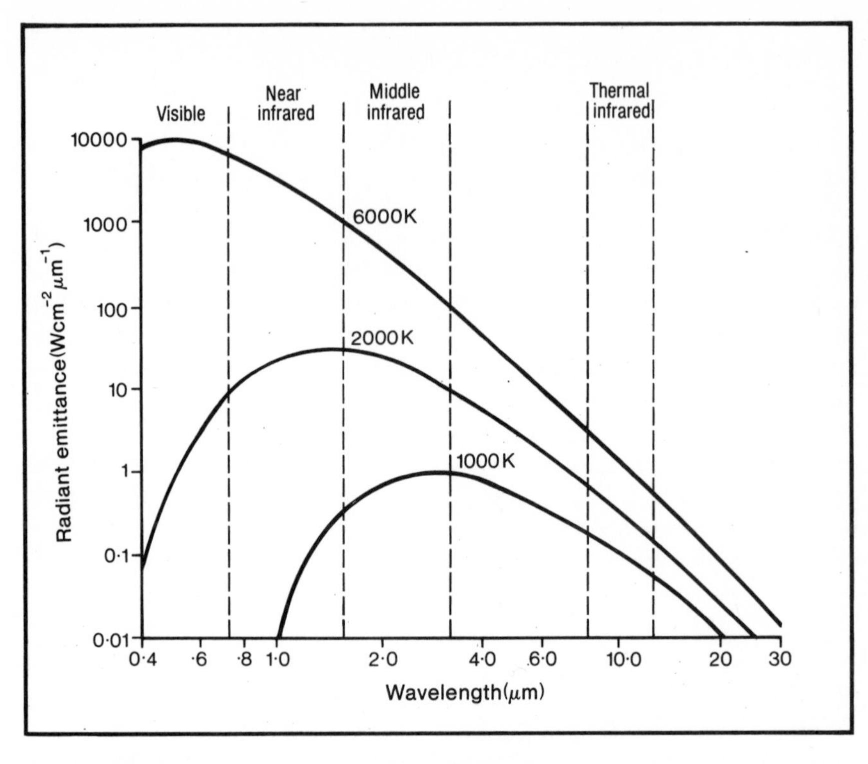


2.2 Relationship between energy and wavelength for three bodies at temperatures of 1000K, 2000K and 6000K. (After Suits 1983)

$$E = \sigma T^4 \tag{2.5}$$

which is the Stefan–Boltzmann law and $\sigma$ is the Stefan–Boltzmann constant ($5.67 \times 10^{-8} Wm^{-2}K^{-4}$).

However, equation 2.5 should be written in full as

$$E = \sigma T^4 \epsilon \tag{2.6}$$

to include the emissivity $\epsilon$. A measurement of E will therefore give a measurement of a combination of T and $\epsilon$. When a blackbody is being measured then $\epsilon = 1$ so can be disregarded, but in some cases (for example, cirrus clouds) $\epsilon$ is less than 1 so its effects must be known before T can be calculated.

## Absorption of radiation by the atmosphere

The simple geometry of satellite remote sensing illustrated in Figure 2.3 shows the relationship between the source of radiation (in this case the sun), the object (the Earth's surface, clouds, etc.) and the sensor on board a satellite. The space between these components is occupied by free space and the Earth's atmosphere, and it is the Earth's atmosphere which is responsible for changing the characteristics of the radiation as it passes from the sun to the Earth and the Earth to the satellite sensor. The Earth's atmosphere absorbs radiation over a wide range of wavelengths because of the large number of gases present in the atmosphere.

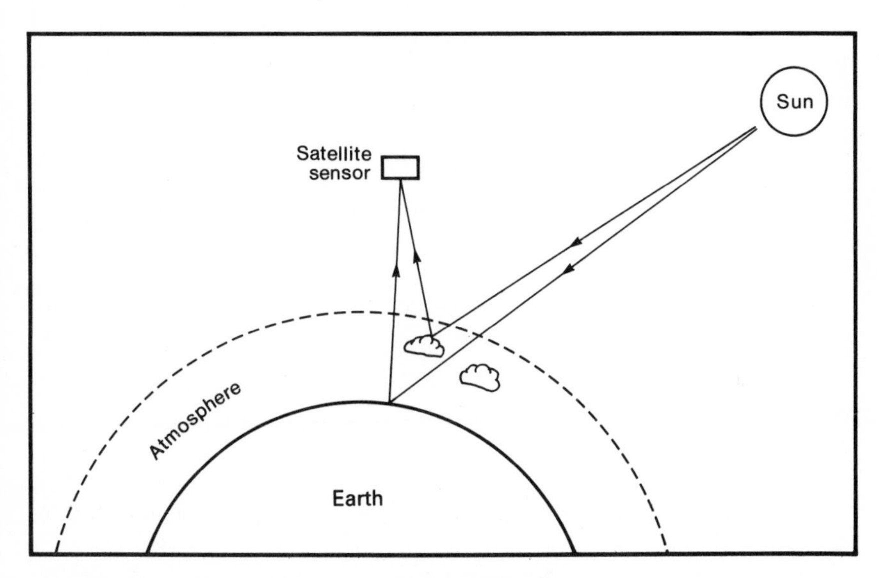


2.3 Simplified geometry of sun, Earth and satellite sensor

Radiation from the sun is shortwave radiation in the wavelength range 0.1–5μm, with a peak at 0.55μm. Radiation emitted by the Earth is longwave radiation principally in the range 5–50μm, with a peak at 10μm. Figure 2.4 shows the spectral distribution of solar and terrestrial radiation. The important absorbing gases in the atmosphere are described below.

*Oxygen and ozone.* Energy of less than 0.1μm wavelength is absorbed in the ionosphere by oxygen molecules ($O_2$) and free oxygen atoms (O). Energy of 0.1–0.3μm and 0.32–0.36μm wavelength is absorbed by ozone ($O_3$).

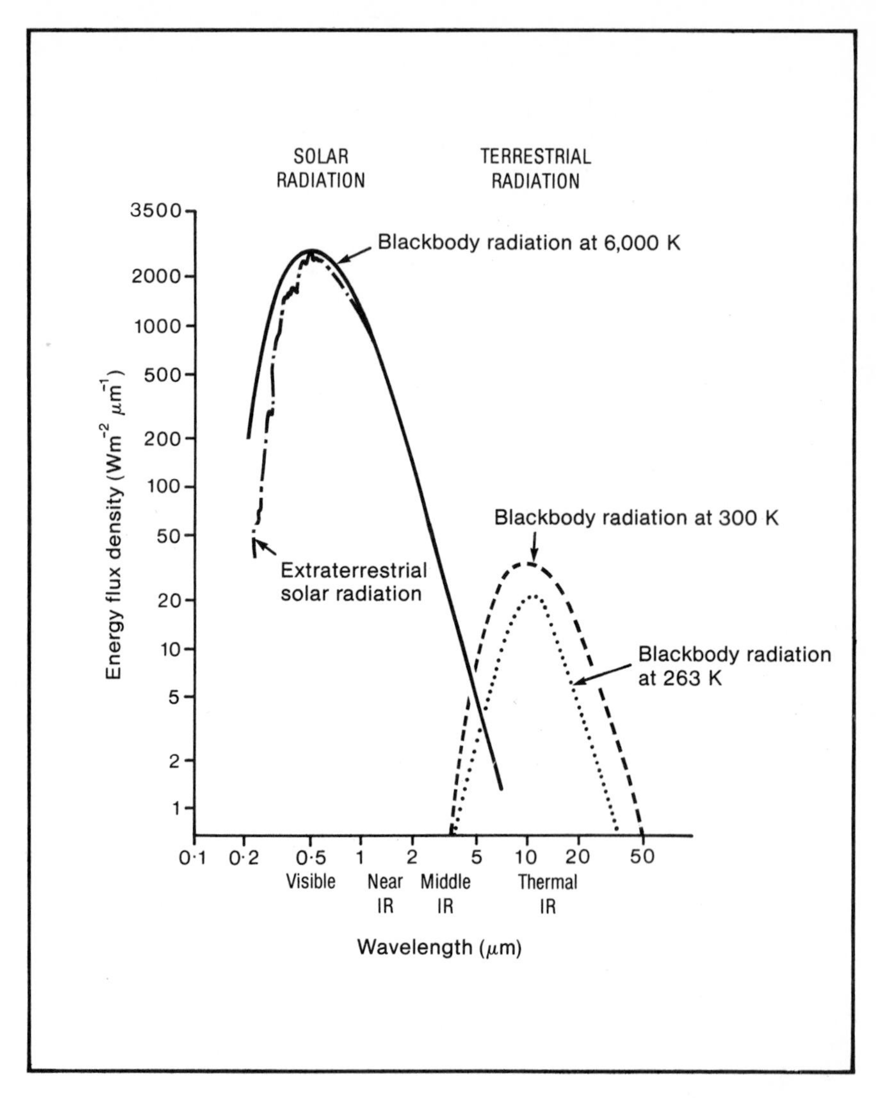


2.4 The spectral distribution of solar and terrestrial radiation. Note the logarithmic scales on both axes. (After Barry and Chorley 1982)

*Carbon dioxide.* There is a strong absorption band at 15µm, and weaker bands at 2.5µm and 4.5µm.

*Water vapour.* There is strong absorption at 6µm and some absorption between 0.6 and 2µm and at 3µm.

As these gases absorb radiation in the wavelengths stated then they modify the characteristics of that radiation, so remote sensing avoids these absorption areas and concentrates on those wavelengths which have low absorption and therefore high transmission. Figure 2.5 gives the atmospheric absorption curves for visible and near infrared and for thermal infrared energy. The diagram shows that the visible, near infrared and thermal infrared regions are *windows* of low absorption (and therefore high transmission) in a spectrum which in other parts shows *high* absorption of radiation. Satellite remote sensing exploits this by concentrating on these atmospheric windows for data collection (see also Table 2.1).

## Reflectance

Electromagnetic energy at short wavelengths ($<5\mu m$) is *reflected* by the surfaces onto which it falls. Energy at longer wavelengths is absorbed then *emitted* by the surfaces onto which energy falls. Energy at microwave wavelengths is *emitted*, or if a radar is used *backscattered*. Backscatter is discussed in chapter 3.

### VEGETATION

The most important distinction to draw in discussing reflectance is between vegetated and non-vegetated surfaces. Vegetation is important because it has a relatively low reflectance in the visible part of the EM spectrum and a relatively high reflectance in the near infrared part (Tucker 1979). Non-vegetated surfaces, for example bare soil, water and concrete, commonly have similar reflectances at visible and near infrared wavelengths. Figure 2.6 shows five simplified reflectance curves for snow, vegetation, soil and water, together with a variety of wavebands used in satellite remote sensing.

At visible wavelengths healthy vegetation has a low blue and red reflectance and a slightly higher green reflectance, hence the green colour of healthy vegetation. In the near infrared the reflectance of vegetation is considerably higher. Stressed or senescent vegetation has a higher visible reflectance and a lower near infrared reflectance than healthy vegetation. The reasons for this pattern are determined by leaf structure.

Figure 2.7 shows a schematic cross-section of a typical citrus leaf. The top layer of cells is the epidermis which diffuses visible light but reflects very little. The palisade cells are long narrow cells holding chloroplasts with chlorophyll pigments which absorb visible light. The spongy mesophyll cells

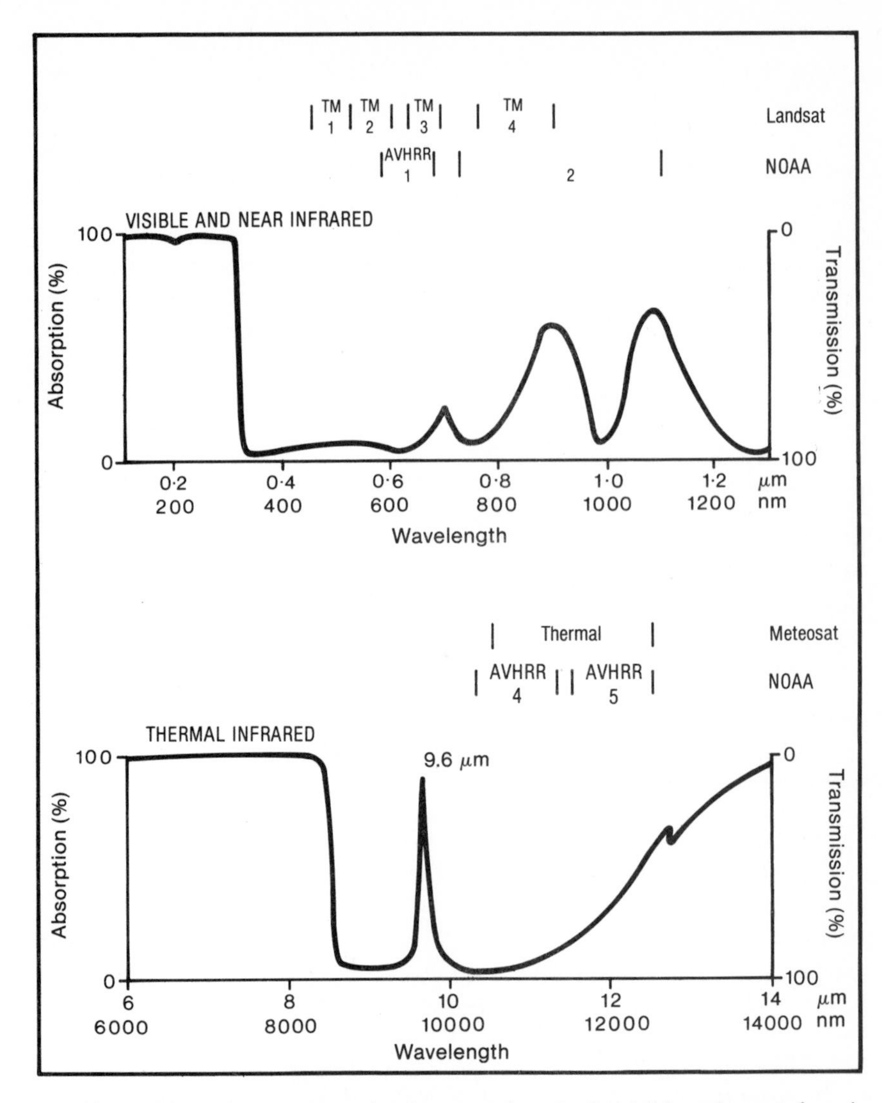


2.5 Absorption and transmission by the atmosphere in the visible and near infrared and in the thermal infrared parts of the electromagnetic spectrum. Example channels for Landsat, Noaa and Meteosat satellite sensors are shown

are very important because they reflect about half of the incoming near infrared radiation. Air spaces in this leaf mesophyll usually develop by the separation of neighbouring cell walls, and it is the increase in the number of air spaces which increases the near infrared reflectance because diffused energy passes more frequently from cells of high refractive index to the

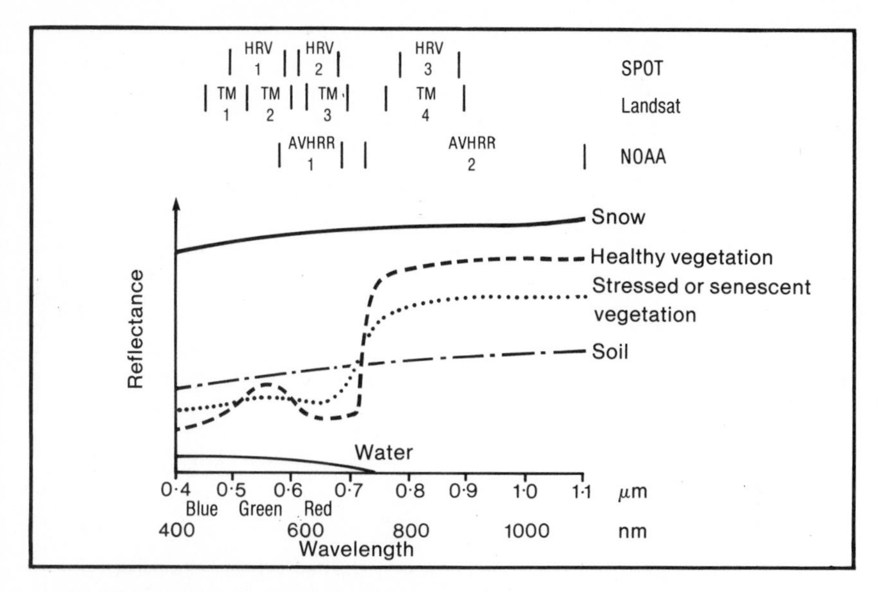


2.6 Simplified spectral reflectance curves for five surfaces in the visible and near infrared parts of the electromagnetic spectrum. Example channels of SPOT, Landsat and Noaa satellite sensors are shown

intercellular air of low refractive index. This leaf structure explains the spectral reflectance curve of healthy vegetation shown in Figure 2.6 and is summarised below (Myers 1983).

Vegetation has a low reflectance in the 0.35–0.7μm region because of absorption by chlorophyll in the palisade cells. There is a slight rise in reflectance at c. 0.55μm, hence healthy leaves appear green, although reflectance is still only about 10% of incoming visible wavelength radiation.

Reflectance of 0.7–1.35μm radiation is higher because of the multiple scattering caused by the leaf's internal spongy mesophyll layer. Reflectance is c. 50% of the incoming near infrared radiation.

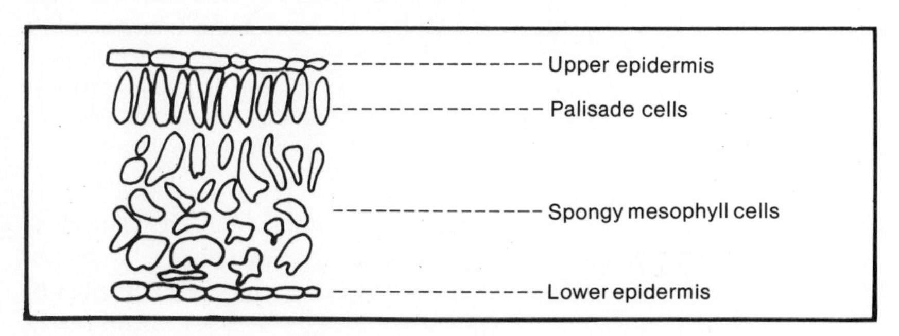


2.7 Schematic cross-section of a typical citrus leaf. (After Myers 1983)

As vegetation becomes stressed or senescent its ability to absorb visible light through chlorophyll absorption in the palisade cells decreases, and as a consequence the amount of visible light reflected increases. Likewise, the spongy mesophyll layer contracts, the density of air spaces decreases and so the reflectance of near infrared radiation also decreases. The increase in visible reflectance and the decrease in near infrared reflectance for stressed vegetation is shown in Figure 2.6.

## NON-VEGETATED SURFACES

The reflectance curves of surfaces other than vegetation normally show smaller changes in reflectance between visible and near infrared wavelengths (see Figure 2.6).

### *Soil*

Soil reflectance depends upon the chemical and physical properties of the soil. Moisture content, organic matter, iron concentration, texture and surface roughness all play a part in determining the spectral reflectance curve (Al Abbas *et al.* 1972, Hoffer 1978, Wright and Birnie 1986). Reflectance of soils gently increases from the visible to the near infrared, and dry soils are normally brighter than wet soils at the same wavelength. Reflectance from soil is important in satellite remote sensing because the reflectance from an area of vegetation is frequently composed of reflectance both from the leaves and from the soil background (Ezra *et al.* 1984).

### *Water*

Water has a higher reflectance at shorter wavelengths (<0.6μm) and virtually no reflectance at wavelengths greater than 0.7μm. At the shorter wavelengths there is some light penetration of water, and this has been used to map water depth in tropical seas with low sediment loads using Landsat data. At wavelengths above 0.7μm water normally has a reflectance close to zero unless the sediment concentration is high (Davies and Charlton 1986).

### *Snow*

Snow surfaces have high reflectance in all visible and near infrared wavelengths. At short wavelengths reflectance increases with wavelength then decreases in the near infrared. Impurities and an increase in grain size reduce reflectance, and consequently as snow ages then reflectance decreases (Smith 1983).

A comparison of visible and near infrared reflectance for vegetated and non-vegetated surfaces can be presented not only as spectral reflectance curves, but also in the form of a two-dimensional distribution as shown in Figure 2.8. This characteristic form shows (1) the plane of soils where the reflectance of bare surfaces (including soils, bare lithology, urban areas) increases simultaneously in both visible and near infrared wavelengths, and (2) the area of vegetation which has a relatively higher near infrared reflectance and a relatively lower visible reflectance. This general form has been referred to as a 'tasseled cap' shape by Kauth and Thomas (1976) because of the cap-like shape of the diagram. The full form of the tasseled cap includes the development of the vegetation reflectance through a growing cycle.

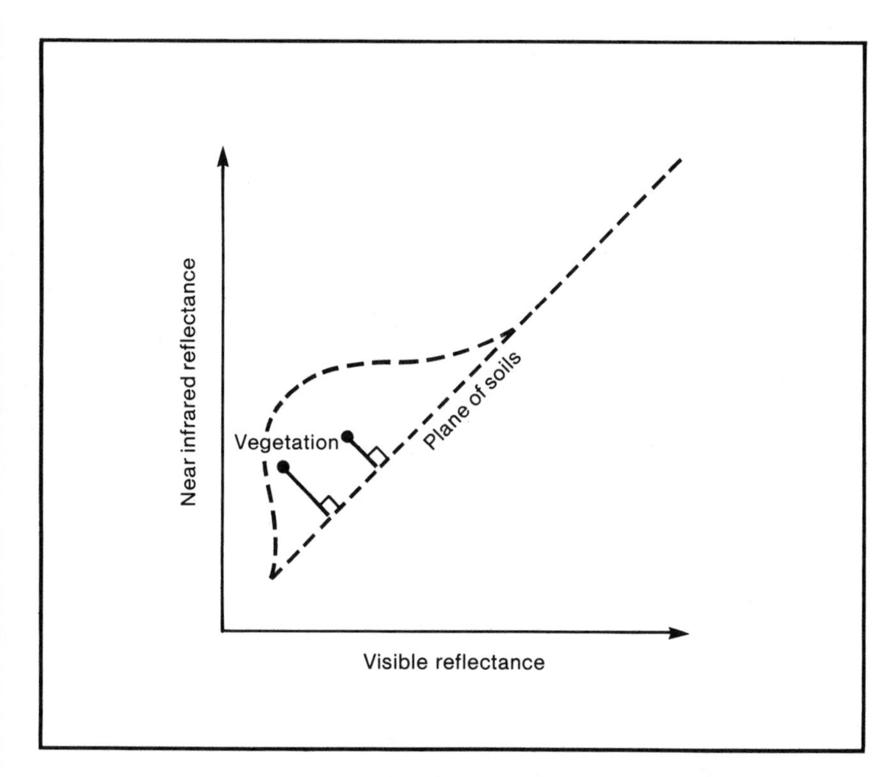


2.8 Simplified relationship of near infrared and visible reflectance showing the characteristic plane of soils and envelope of vegetation where visible reflectance is relatively low and near infrared reflectance relatively high. This shape has been termed the 'tasseled cap' (Kauth and Thomas 1976). The lines plotted at right angles to the plane of soils are to illustrate the Perpendicular Vegetation Index which is discussed in chapter 6

# *Chapter 3*

# Satellite sensor systems

Radiation reflected by or emitted from surfaces is collected by some form of sensor on board a satellite platform. The sensor systems employed in satellite remote sensing include photographic equipment, television cameras, scanning radiometers and imaging radars, and these systems are discussed in this chapter.

## Photographic systems

Photographic camera systems in space use precision photogrammetric technology developed originally for aerial photographic missions, but raised to the higher altitude of a satellite orbit to allow a larger coverage from a stable platform.

The manned space missions of the US Gemini and Apollo series of the 1960s produced many photographs of parts of the Earth taken by the astronauts, and these photographs revealed both the beauty and the scientific interest of the Earth viewed from space. The first serious collection by NASA of multiband photographs of the Earth from space was the Apollo 9 mission of March 1969 which carried a set of four Hasselblad cameras, the S095 experiment, attached to the Apollo 9 command module. Each of the four cameras had a different filter so that multispectral photographs of the Earth could be obtained.

An expanded experiment consisting of six cameras was carried as the S190A experiment on board the Skylab satellite which was launched in May 1973. The S190A experiment had four cameras in a multispectral combination operating in the range 0.5–0.9μm, and two cameras operating at 0.5–0.88μm and 0.4–0.7μm. Skylab also carried the S190B experiment which had just one waveband (0.4–0.9μm) and a high spatial resolution of 20m. The S190B experiment was also termed the Earth Terrain Camera, and was a high performance reconnaissance camera incorporating forward motion compensation. The S190A and S190B experiments together with hand-held 35mm and 70mm format cameras were used to take over 35,000 photographs of the Earth (Curran 1985). The Soviet Soyuz satellites have also carried cameras for taking photographs of the Earth, and Soyuz 22 was equipped with a Zeiss Jena MKF-6 camera with six lenses for multispectral operation.

In the first half of the 1980s the NASA Space Shuttle provided a suitable platform for photography from space. In 1983 the Spacelab 1 module on the Space Shuttle carried the European Space Agency's Metric Camera experiment. The camera was a Zeiss RMK A30/23 modified for operation in space. The coverage of each Metric Camera frame was 190km × 190km, and as each film frame measured 23cm × 23cm then this produced a film scale of 1:826,000 and a ground resolution of 20m. Enlargement to scales of 1:100,000 to 1:50,000 has been possible, although the film quality was poorer than anticipated because the Shuttle mission was flown in November 1983 instead of August 1983, so light levels in the northern hemisphere were lower than anticipated when the mission was planned.

The NASA counterpart to ESA's Metric Camera is the Large Format Camera (LFC) which was flown on the Space Shuttle in 1984. The film size is 23cm × 46cm, twice the area of the Metric Camera, and the ground resolution is of the order of 20m (Anon 1985, Doyle 1985). The Large Format Camera has forward image motion compensation so that during exposure the platen moves the film at exactly the correct speed to cancel the image smear which would otherwise be caused by the movement of the Space Shuttle over the Earth's surface. Image overlap of up to 80 per cent was acquired with the LFC so that at a scale of 1:50,000 a height accuracy of 30m could be achieved. In areas of high tonal contrast ground resolutions up to 5m have been noted (Togliatti 1986).

## Television cameras

The first weather satellite, Tiros 1 launched in 1960, carried a television camera as its primary sensor. Television cameras continued to be used on the early weather satellites, and later on Landsat, but they suffer from problems of radiometric fidelity. The Landsat television camera was a Return Beam Vidicon (RBV). On Landsats 1 and 2 three RBV cameras were used, each corresponding to a different waveband:

| | | |
|---|---|---|
| RBV1 | 0.475–0.585μm | green |
| RBV2 | 0.58 –0.69μm | red |
| RBV3 | 0.69 –0.83μm | near infrared |

Power problems caused early failure of the RBVs and on the whole the results from these sensors were disappointing. Landsat 3 carried a single RBV with one broad waveband of 0.51–0.75μm and a spatial resolution of c. 30m. Figure 3.1 shows a segment of a Landsat 3 RBV image of Kuwait and illustrates the high topographic detail of this sensor. Roads, urban areas and coastal features are clearly revealed. The white patches on the right of the image are small cumulus clouds with accompanying dark shadows to their upper left (north-west) side. The cross marks on Figure 3.1 are reseau marks used for geometric rectification. Even the Landsat 3 RBV was disappointing because of its low radiometric resolution: the imagery are useful for visual interpretation but less so for accurate quantitative applications.

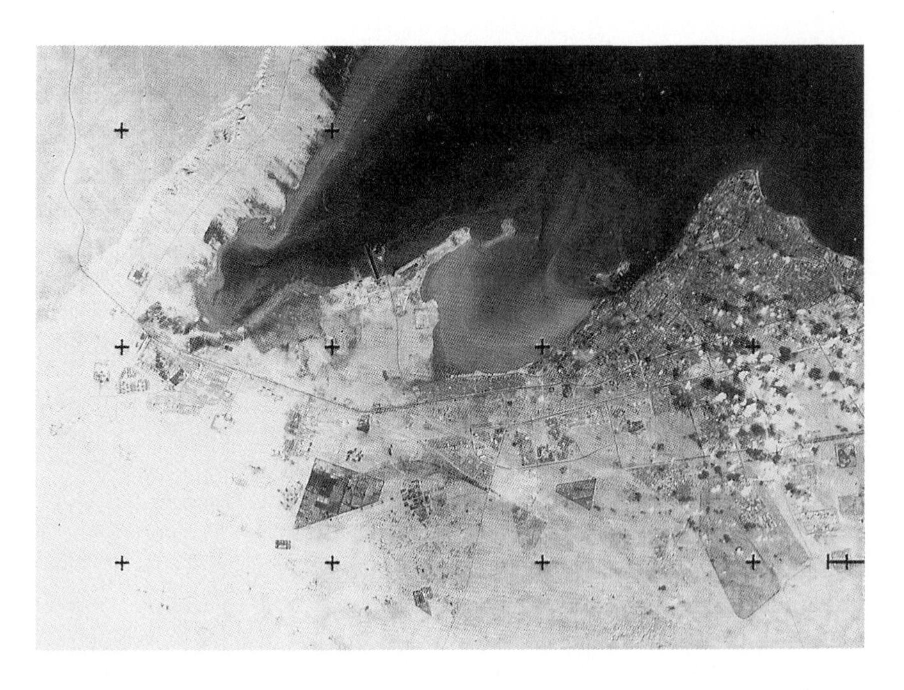

3.1 Segment of a Landsat 3 Return Beam Vidicon (RBV) image of Kuwait, 29 November 1980. The crosses are reseau marks. The white patches on the right (east) of the image are small cumulus clouds

## Scanning radiometers

The primary instrument used for satellite remote sensing in the visible and the infrared is the scanning radiometer. The basic form of this instrument consists of a mirror system to focus EM energy from the Earth onto a detector, and a scanning device to vary the origin of that energy. Figure 3.2 shows the principle elements of a scanning radiometer in diagrammatic form. The scanning radiometer operates by focussing EM energy onto a detector. This energy arrives at the detector from an oscillating mirror which scans the surface below the satellite along a line perpendicular to the direction of the satellite's orbit (see Figure 3.2). The detector section of the scanning radiometer can be built to divide the incoming energy into a number of different wavebands, say visible, near infrared and thermal infrared. As the scanner traverses a scan line a voltage is produced at the detector which is a function of the amount of incoming EM energy. This voltage is sampled at regular intervals and converted to digital form. This sampling in time is the equivalent of sampling geographically along a scan line. Each sample along a scan line is the energy collected by an instantaneous field of view (IFOV) of the sensor and is referred to as a pixel. A single scan line comprises a line of pixels. At the end of each scan the mirror views a set of calibration points commonly in the form of a grey wedge and other standard references. This allows calibration of the sensor signal from one scan to the next and prevents the sensor response changing through time.

A satellite image from a scanning radiometer is constructed from a series of adjacent scan lines, each scan line being a line of pixels. Successive scan lines are added as the orbital motion of the satellite allows a fresh scan of a new line to be taken after the previous oscillation of the scanning mirror has been completed (see Figure 3.2).

Scanning radiometers have been carried on board a variety of remote sensing satellites including the Landsat Multispectral Scanner (MSS) and Thematic Mapper (TM), the Noaa Advanced Very High Resolution Radiometer (AVHRR) and the Nimbus Coastal Zone Colour Scanner (CZCS). Scanning radiometers in a modified form are also the main imaging sensors on the geostationary weather satellites, including the US GOES, Europe's Meteosat and Japan's GMS.

## Imaging radar

Imaging radars on satellites are at present in an experimental phase, and

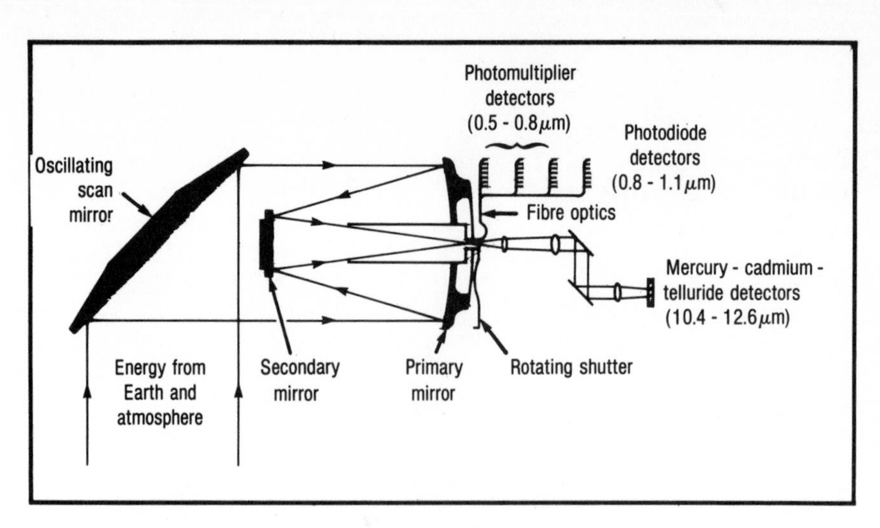


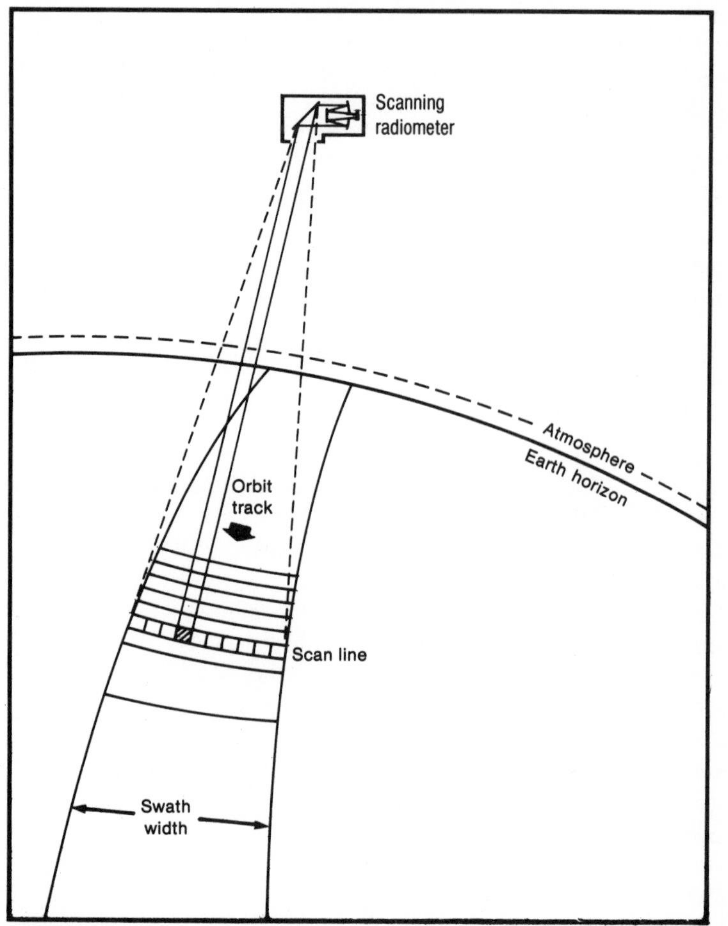


3.2 Schematic diagram of the Landsat 3 Multispectral Scanner scanning radiometer (top) and its orbital scanning pattern. (In part after Freden and Gordon 1983)

thus far only limited data from Seasat and from the two Shuttle Imaging Radars (SIR-A and SIR-B) have been collected. However, in the 1990s radar remote sensing will be a significant satellite remote sensing tool and satellites planned for launch around 1990 include ESA's ERS-1, Canada's Radarsat and Japan's J-ERS-1. In order to discuss satellite imaging radar, which effectively has to be synthetic aperture radar, it is necessary first to discuss the experience gained with real aperture radars carried on board aircraft.

Radar remote sensing uses entirely different wavelengths than visible and infrared remote sensing. Radar employs EM energy in the microwave part of the spectrum, and radar wavelengths are measured in mm or cm. Table 3.1 gives some of the commonly used wavelengths together with their letter designations and frequencies. Because radar wavelengths are so long they are not affected by the relatively small cloud particles and radar can therefore penetrate clouds and as it were 'see' the ground surface in cloudy conditions. This is of great benefit in the cloudy mid-latitude and tropical areas.

*TABLE 3.1: Wavebands commonly used in radar remote sensing*

| *Band designation* | *Wavelength range (mm)* | *Frequency (GHz)* |
|---|---|---|
| Ka | 8–11 | 28–37.5 |
| K | 11–17 | 18–28 |
| X | 24–38 | 8–12.5 |
| C | 38–75 | 4–8 |
| L | 150–300 | 1–2 |

In addition to its wavelength characteristics, radar energy can also be polarised. Polarisation indicates the plane in which the EM energy travels. Radar systems employ horizontal (H) and vertical (V) polarisation in the combinations shown in Table 3.2.

*TABLE 3.2: Polarisation combinations for radar*

| *Transmitted energy* | | *Received energy* | |
|---|---|---|---|
| Horizontal | HH | Horizontal | Like polarisation |
| Vertical | VV | Vertical | Like polarisation |
| Horizontal | HV | Vertical | Cross polarisation |
| Vertical | VH | Horizontal | Cross polarisation |

Side-looking airborne radar (SLAR) uses real aperture radar for imaging applications. A SLAR radar antenna is carried on board an aircraft and sends out to one side of the aircraft a pulse or beam of EM energy in the wavelength chosen for the radar (see Table 3.1) at right angles to the direction of flight of the aircraft. The energy is *backscattered* from the surfaces it hits and is partially returned to the SLAR antenna which measures the energy returned (Hoekman 1985). Figure 3.3 illustrates the geometry of a single pulse. The SLAR antenna is capable of operating in either transmit or receive mode, and in order to send out pulses of radar energy and receive the backscattered energy the antenna is switched electrically between the two modes. An image is constructed by adjusting the antenna monitor so that the successive pulses, and their corresponding backscatter return, produce a series of adjacent lines along the flight line of the aircraft. These lines then make up the SLAR image.

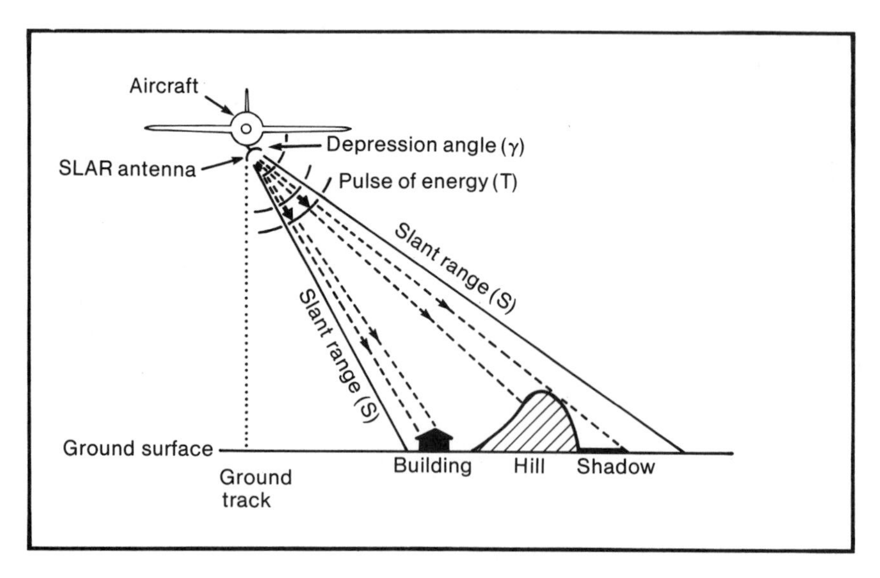


3.3 Schematic illustration of a side-looking airborne radar (SLAR)

The process described above is one of active illumination at microwave wavelengths and so can be operated day or night independently of solar illumination.

The amount of energy returned or backscattered to the antenna from the surface depends upon two main factors:

(1) The angle of the surface to the incident radar beam. Backscatter will be high from those surfaces at right angles to the beam, such as the left

side of the house roof in Figure 3.3; low from those surfaces which slope away from the radar beam, such as the right side of the house roof in Figure 3.3; and non-existent (i.e. shadow) in those areas hidden from the active illumination of the radar pulse, such as the right side of the hill in Figure 3.3.
(2) The physical properties of the surface, primarily its surface roughness and its dielectric constant. Smooth surfaces backscatter little or no radar energy while rough surfaces backscatter more energy depending upon their surface roughness. The surface dielectric constant is largely determined by water content. Wet soil, for example, will have a high dielectric constant and a high backscatter; dry sand will have a low dielectric constant and a lower backscatter.

### *Range resolution*

In the case of SLAR resolution improves with range, so that objects further from the ground track of the aircraft can be more easily separated than objects closer to the ground track. Resolution in range $R_r$ is governed by the equation

$$R_r = \frac{Tc}{2 \cos \gamma} \tag{3.1}$$

where T is the duration of the radar pulse (pulse length), c is the speed of light, and $\gamma$ is the depression angle (see Figure 3.3).

### *Azimuth resolution*

In contrast to range resolution, the azimuth resolution, that is the resolution in the along track direction, deteriorates with distance from the aircraft ground track. The equation for the azimuth resolution ($R_a$) is

$$R_a = \frac{0.7 \, S \, \lambda}{D} \tag{3.2}$$

where S is the slant range, $\lambda$ is the wavelength, and D is the length of the antenna. As $\lambda$ and D are constant for any single SLAR configuration then it is clear from equation 3.2 that the azimuth resolution depends upon S, the slant range. Figure 3.3 shows that S is greatest at the outer limb of the pulse of radar energy. Another way of looking at the azimuth resolution is to envisage the beam of radar energy broadening away from the aircraft so that at some distance away the beam is wider than close to the ground track. As the beam becomes larger away from the aircraft ground track then its ability to resolve separate objects becomes poorer.

## SATELLITE RADAR

The equation for azimuth resolution (equation 3.2) illustrates some of the problems of extending SLAR to satellite altitudes. The slant range S would increase from an aircraft altitude of (say) 5km to a satellite altitude of (say) 700km. This would dramatically increase the value of $R_a$ and make the azimuth resolution poorer. In equation 3.2 there are two ways to combat this; either decrease the operating wavelength $\lambda$ or increase the antenna length D. A few quick calculations of equation 3.2 with some sample values for S, $\lambda$ and D will show this. A decrease of $\lambda$ is impractical because wavelengths shorter than c. 8mm are affected by rain and cloud droplets. An increase in D is impractical because a very large antenna (kilometres in length) would be required which would be costly, power-consuming, difficult to construct in space, and subject to instability. It is possible that long antennae could be constructed on an orbiting space station in the twenty-first century, but for the present an alternative answer must be sought.

The answer to the problem has come by manipulating the radar pulse returns to synthesise a long antenna, and hence increase D, but still employ a short physical antenna. This is termed synthetic aperture radar (SAR) because a long aperture (antenna) is synthesised from a series of small apertures.

The synthetic aperture concept may be thought of in terms of an object on the ground and successive pulses of radar energy. The situation is shown diagrammatically in Figure 3.4. With a wide beam from an antenna then the object falls within successive pulses of energy and stays within the active illumination of the radar for a longer period than is the case with a single narrow beam. If the returns from the ground object for the successive beams are stored by the radar then combined in the correct phase, then a narrow, high resolution radar can be synthesised from a series of wide beam pulses from successive antenna positions (Moore 1983). In fact for a SAR system $R_a$ reduces to

$$R_a = \frac{D}{2} \tag{3.3}$$

and so becomes independent of satellite altitude.

Synthetic aperture radar offers a number of important advantages for satellite remote sensing, but there are also a number of disadvantages associated with its use.

*Advantages*

(1) Active illumination of the target, so SAR is independent of solar illumination.

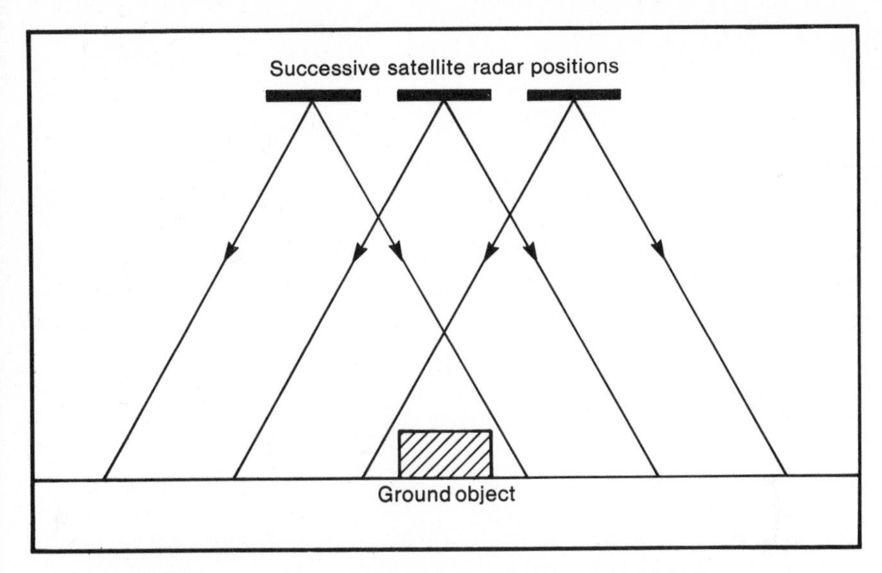


3.4 Synthetic aperture radar geometry. The ground object stays within the field of view of the radar beam for three successive pulses

(2) Signal is not affected by cloud, so SAR is independent of weather and can 'see' the surface through clouds.
(3) Provides oblique illumination.
(4) Can record information at different polarisations.

*Disadvantages*

(1) The processing of SAR data is very complex and time-consuming, although near real time processing is now feasible at specialist installations.
(2) Shadows occur where ground objects obstruct the passage of a radar pulse, and so information is lost.
(3) Resolution is variable with distance from the ground track, although this can be corrected by resampling.
(4) In mountainous areas serious geometric distortions in the imagery are created.

Figure 3.5 shows an example of an image produced by a satellite-borne synthetic aperture radar. It is a Shuttle Imaging Radar (SIR-A) image of a part of central Tunisia south of Sousse (Harris 1985b). The image shows the urban areas as bright patches particularly well because of their high backscatter, and also shows a clear land/sea contrast, the presence of smooth inland sebkhas or playa lakes, and the variation in local agriculture.

3.5 Shuttle Imaging Radar (SIR-A) L-band image of central Tunisia. (Courtesy Jet Propulsion Laboratory, California)

## New sensor systems

One problem with the scanning radiometer is that it is a mechanical system: the oscillating mirror is driven by a motor. This can lead to problems of varying speed and to problems of mechanical failure. One way of overcoming these problems is to replace the scanning mirror system, which collects radiance data from one pixel at a time, with a solid state multilinear array (MLA). The MLA is a line of sensors, each capable of collecting radiance data simultaneously from neighbouring pixels along a line at right angles to the direction of flight of the satellite. Each sensor element of an MLA is a charge-coupled device (CCD). Table 3.3 gives the comparative advantages and disadvantages of the MLA over conventional mechanical scanning radiometers.

The SPOT satellite is equipped with an MLA. The basic detector of the SPOT MLA has 1,728 cells across so that it can collect data for 1,728 pixels simultaneously by electronic control of the exposure of each MLA element, yet the MLA detector only measures 80mm in length.

The MLA is also referred to as a pushbroom scanner because the linear

*TABLE 3.3: Advantages and disadvantages of multi-linear arrays*

| *Advantages of multi-linear arrays over mechanical scanning radiometers* | *Disadvantages of multi-linear arrays compared with mechanical scanning radiometers* |
|---|---|
| Lighter weight | Many more detectors to calibrate |
| Smaller size | Cannot sense wavelengths longer than near infrared |
| Lower power requirements | |
| No moving parts | |
| Longer life expectancy | |
| Greater reliability | |
| Higher geometric accuracy | |
| Higher radiometric accuracy | |
| Higher signal to noise ratio | |
| Higher spatial resolution | |
| Lower cost | |

*Source*: Curran (1985)

array is compared to a broom head which is pushed along the satellite flight line, collecting radiance data as it does so. Because of the lack of moving parts the MLA on SPOT should be more reliable than scanning radiometers, and more satellites in the near future, for example the Japanese MOS-1, will be equipped with these new sensors.

## Resolution

In the discussion of sensor systems in this chapter the resolution of the system is of considerable importance when assessing its utility. Two elements of resolution are important: spatial resolution and spectral resolution.

Spatial resolution is commonly quoted as the pixel size, which in turn is a function of the instantaneous field of view (IFOV) of the sensor system. This is particularly common for scanning radiometers. Thus the Landsat Thematic Mapper is quoted as having a spatial resolution of 30m, and the SPOT High Resolution Visible sensor a spatial resolution of 20m. However, the spatial resolution of a sensor system is not only governed by the pixel size, but also by a number of other factors including (Forshaw *et al.* 1983):

Target geometry and contrast
Radiometric resolution
Illumination
Clarity of the atmosphere
Effects of neighbouring pixels

Thus the spatial resolution can be strongly influenced by the contents of a scene: a sand desert of low contrast imaged by the Landsat Thematic Mapper will have an effective spatial resolution (that is equivalent pixel size) of greater than 30m, while a narrow tarmac road across that sand desert will be visible because of its brightness contrast even though the width of the road itself may be less than 30m. See, for example, Alwash *et al.* (1986) and Latham and Hubbard (1985).

The spectral resolution concerns the width of the waveband used in the sensor system. As satellite remote sensing has developed then systems have been designed with narrower wavebands. For example, the first three bands of the Landsat Multispectral Scanner each have a bandwidth of 0.1μm, while its successor, the Thematic Mapper, has for its first three wavebands bandwidths of 0.07, 0.08 and 0.06μm respectively. This improvement in spectral resolution is partly the result of improvements in our knowledge of the reflectivity characteristics of terrain, and a more precise definition of the objectives of satellite missions (NOAA 1986).

*Chapter 4*

# Ground data collection

One important element of data acquisition in remote sensing is ground or *in situ* data collection. This element of remote sensing is important for the interpretation of satellite remote sensing data both qualitatively and quantitatively. This chapter discusses the information needs for *in situ* data collection for a range of wavelengths.

The discussion in this chapter concentrates on measurements of the land surface. Many of the considerations are the same for *in situ* measurements of water, ice and snow and of the atmosphere, but some of the observations will be different. Meteorologists use the network of conventional meteorological observing stations to provide *in situ* information on temperature, cloud cover, wind direction, etc. Oceanographers measure surface reflectance and backscatter as described in this chapter, and also measure variables such as salinity, turbidity and chlorophyll concentration (see chapter 10).

The type of investigation and the time scale are of related importance. A distinction must be made between environmental features which are stable through time and those which change rapidly. In the case of the former then the ground data collection and the satellite data collection do not have to be coincident: indeed there may be months or years between the two. This may be the case in geology where rock type and structure will not change over several months, unless a catastrophe such as an earthquake or a volcanic eruption occurs. In the case of features which do change rapidly then the timing of the ground data collection in relation to the satellite overpass is critical. For example, soil moisture changes on an hourly scale, so *in situ* measurements of soil moisture must be as close as possible to the time of the satellite data collection. An illustration of an intermediate time scale is agriculture, where ground data collection within days of the satellite overpass is normally acceptable, unless of course the crop has been harvested in the meantime.

## Visible and near infrared

### QUALITATIVE

The recording of surface characteristics in the field at selected sites provides a record of surface conditions which can be related to satellite remote sensing data. Table 4.1 gives examples of the types of field records collected for four environmental applications: geology, agriculture, urban and water. Information collected about these attributes on the ground at known points allows the correct interpretation of satellite remote sensing data. This interpretation can take two forms:

(1) Visual interpretation of satellite imagery.
(2) Categorical analysis of the radiometric characteristics of pixels at the ground site locations. This amounts to classifier training and is discussed later in chapter 6.

Analysis of the satellite data can then proceed by working out from the known ground areas to those areas not visited in the field yet imaged by the satellite.

*TABLE 4.1: Environmental characteristics recorded in the field for four applications of satellite remote sensing*

| *Geology* | *Agriculture* | *Urban* | *Water* |
|---|---|---|---|
| Rock type and lithology | Field size | Housing density | Colour |
| Structure | Primary species | Housing pattern | Movement |
| Lineaments | Irrigation | Building materials | Pattern |
| Outcrops | Row direction | Sizes | Width |
| Colour | Row width | Surface colour | Depth |
| Soil type | Crop height | Road sizes | Salinity |
| | Crop colour | Road surfaces | Turbidity |
| | Ground cover % | Vegetation | Secchi disk depth |
| | Soil type | | |
| | Slope | | |

## QUANTITATIVE

The second form of ground data collection is the use of a ground radiometer for the collection of quantitative information about the surface. This instrument is used to measure the reflectance of the surface in wavelengths which are comparable to those used in the satellite sensor. An example of a ground radiometer is a Milton Multiband Radiometer (Milton 1980), and one of these instruments is illustrated in use in Figure 4.1. The radiometer's sensor head is fixed to a stand 2m above the ground surface and points downwards. The reflectance from the surface is measured by detectors in the radiometer, each of which has a different filter to detect a different part of the EM spectrum. In addition to the reflectance from the ground surface, the reflectance of a standard diffuse reflectance surface, such as barium sulphate or magnesium oxide, is also recorded. In Figure 4.1 this standard reflectance surface is shown as a grey card. This allows the calculation of a standard measure of reflectance, the bidirectional reflectance:

$$B = \frac{R_g}{R_s} K_\lambda \qquad (4.1)$$

where B is the bidirectional reflectance, $R_g$ is the reflectance from the ground surface, $R_s$ is the reflectance from the standard surface and $K_\lambda$ is a wavelength-dependent constant, measured from laboratory calibration of the standard reflectance surface. As an illustration, from calibrations performed at the University of Southampton using a barium sulphate Kodak grey card as the diffuse reflectance surface $K_\lambda$=20.95 for Landsat Thematic Mapper band 2 equivalent (0.52–0.60μm) and $K_\lambda$=25.15 for Landsat Thematic Mapper band 4 equivalent (0.76–0.90μm) (E.J. Milton personal communication). Silva (1978) discusses in detail the various measures of and terminology for reflectance.

An example of the bidirectional reflectance in the visible and near infrared is shown in Figure 4.2 for surfaces in central Tunisia. The data were collected using a Milton Multiband Radiometer in an area south-east of Sidi Ali ben Aoun (see Table 4.2). Figure 4.2 illustrates the plane of soils and the vegetation concentration discussed in chapter 2 (see Figure 2.8 for comparison). The bare sandy and stony soil reflectances lie on the diagonal, while the measurements of the cereal vegetation lie to the upper left of this plane of soils. The presence of vegetation decreases the visible reflectance (x axis – bidirectional reflectance for Thematic Mapper band 2 equivalent) because of chlorophyll absorption, and increases the near infrared reflectance (y axis – bidirectional reflectance for Thematic Mapper band 4 equivalent) because of the intercellular spaces in the leaf structure.

The wavebands used in ground radiometers are made comparable to

4.1 A Milton Multiband Radiometer in use

*TABLE 4.2: A recording form for data collection using a four-channel ground radiometer. The example data shown were collected in an area south-east of Sidi Ali ben Aoun in central Tunisia, and were used to construct part of Figure 4.2. Three readings are taken at each site and the mean values calculated. Type* A *is sand with low shrubs; type* B *is barley. (Data courtesy Brian Harris and Sinclair Sutherland)*

| *te* | *Time* | *Site number* | *Site type* | *Ground surface* Channel* | | | | *Standard reflectance surface* Channel* | | | |
|---|---|---|---|---|---|---|---|---|---|---|---|
| | | | | 1 | 2 | 3 | 4 | 1 | 2 | 3 | 4 |
| 1/85 | 1120 | 1.1 | *A* | 0.930 | 0.194 | 0.337 | 0.060 | 0.127 | 0.176 | 0.233 | 0.040 |
| 1/85 | 1122 | 1.2 | *A* | 0.940 | 0.195 | 0.340 | 0.060 | 0.130 | 0.180 | 0.239 | 0.041 |
| 1/85 | 1123 | 1.3 | *A* | 0.096 | 0.199 | 0.341 | 0.061 | 0.125 | 0.174 | 0.231 | 0.039 |
| | *Means at site 1* | | | 0.943 | 0.196 | 0.339 | 0.060 | 0.127 | 0.177 | 0.234 | 0.040 |
| 1/85 | 1125 | 2.1 | *A* | 0.103 | 0.215 | 0.358 | 0.066 | 0.122 | 0.170 | 0.225 | 0.038 |
| 1/85 | 1127 | 2.2 | *A* | 0.105 | 0.218 | 0.363 | 0.068 | 0.126 | 0.175 | 0.232 | 0.039 |
| 1/85 | 1129 | 2.3 | *A* | 0.106 | 0.220 | 0.364 | 0.068 | 0.126 | 0.176 | 0.233 | 0.040 |
| | *Means at site 2* | | | 0.105 | 0.218 | 0.362 | 0.067 | 0.125 | 0.174 | 0.230 | 0.039 |
| 1/85 | 1132 | 3.1 | *A* | 0.090 | 0.185 | 0.334 | 0.059 | 0.130 | 0.181 | 0.240 | 0.041 |
| 1/85 | 1133 | 3.2 | *A* | 0.092 | 0.188 | 0.337 | 0.059 | 0.129 | 0.180 | 0.237 | 0.041 |
| 1/85 | 1135 | 3.3 | *A* | 0.092 | 0.189 | 0.338 | 0.060 | 0.131 | 0.183 | 0.242 | 0.042 |
| | *Means at site 3* | | | 0.091 | 0.187 | 0.336 | 0.059 | 0.130 | 0.181 | 0.240 | 0.041 |
| | . | . | | . | | | . | | | | |
| | . | . | | . | | | . | | | | |
| | . | . | | . | | | . | | | | |
| 1/85 | 1200 | 9.1 | B | 0.071 | 0.133 | 0.393 | 0.067 | 0.139 | 0.194 | 0.256 | 0.045 |
| 1/85 | 1202 | 9.2 | B | 0.072 | 0.135 | 0.396 | 0.068 | 0.140 | 0.195 | 0.257 | 0.045 |
| 1/85 | 1203 | 9.3 | B | 0.072 | 0.134 | 0.395 | 0.068 | 0.138 | 0.192 | 0.253 | 0.044 |
| | *Means at site 9* | | | 0.072 | 0.134 | 0.395 | 0.068 | 0.139 | 0.194 | 0.255 | 0.045 |
| 1/85 | 1205 | 10.1 | B | 0.061 | 0.131 | 0.345 | 0.058 | 0.140 | 0.195 | 0.256 | 0.045 |
| 1/85 | 1207 | 10.2 | B | 0.067 | 0.134 | 0.348 | 0.058 | 0.141 | 0.196 | 0.258 | 0.045 |
| 1/85 | 1209 | 10.3 | B | 0.066 | 0.133 | 0.346 | 0.058 | 0.134 | 0.187 | 0.246 | 0.043 |
| | *Means at site 10* | | | 0.065 | 0.133 | 0.346 | 0.058 | 0.138 | 0.193 | 0.253 | 0.044 |
| 1/85 | 1213 | 11.1 | B | 0.070 | 0.135 | 0.349 | 0.058 | 0.141 | 0.196 | 0.259 | 0.045 |
| 1/85 | 1215 | 11.2 | B | 0.070 | 0.136 | 0.350 | 0.057 | 0.138 | 0.192 | 0.254 | 0.044 |
| 1/85 | 1216 | 11.3 | B | 0.070 | 0.135 | 0.354 | 0.058 | 0.139 | 0.193 | 0.255 | 0.045 |
| | *Means at site 11* | | | 0.070 | 0.135 | 0.351 | 0.058 | 0.139 | 0.194 | 0.256 | 0.045 |
| | . | . | | . | | | . | | | | |
| | . | . | | . | | | . | | | | |
| | . | . | | . | | | . | | | | |

The channels are equivalent to Landsat Thematic Mapper bands 2, 3, 4 and 5 (see Table 5.7)

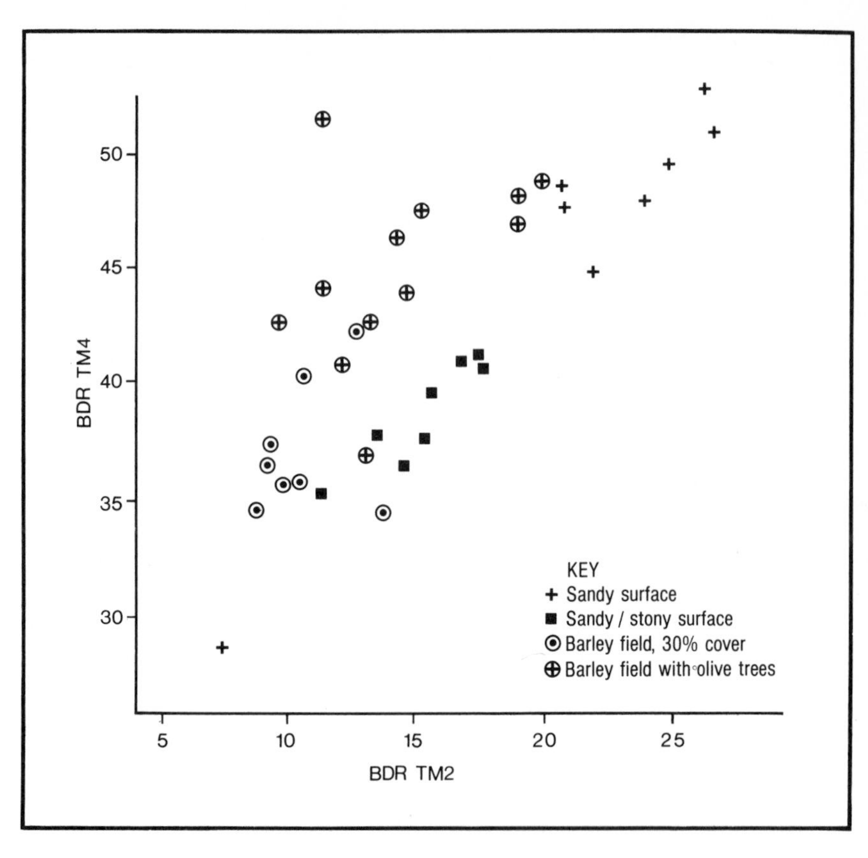


4.2 Bidirectional reflectance (BDR) for four surface types in the Sidi Ali ben Aoun region of central Tunisia. The data were collected using a Milton Multiband Radiometer. The two axes are: BDR TM2 0.52–0.60μm; BDR TM4 0.76–0.90μm

satellite sensor wavebands by choosing filters whose spectral reflectance curves approximate those of the satellite sensor. Normally this is achieved by taking the 50 per cent sensitivity point on the lower and upper limbs of the response curve for the filter/detector combination employed in the ground radiometer. This then gives the waveband interval. An example is shown in Figure 4.3. This is a near infrared band with a waveband interval of 0.78–1.02μm as identified by the points of 50 per cent relative response on the curve. This range was designed to be comparable to the 0.8–1.1μm range of the Landsat Multispectral Scanner band 7 detector.

The quantitative comparison of ground radiometer data with satellite data from scanning radiometers, multi-linear arrays, etc. is a difficult process, so most studies have compared the results of the two data sets in a

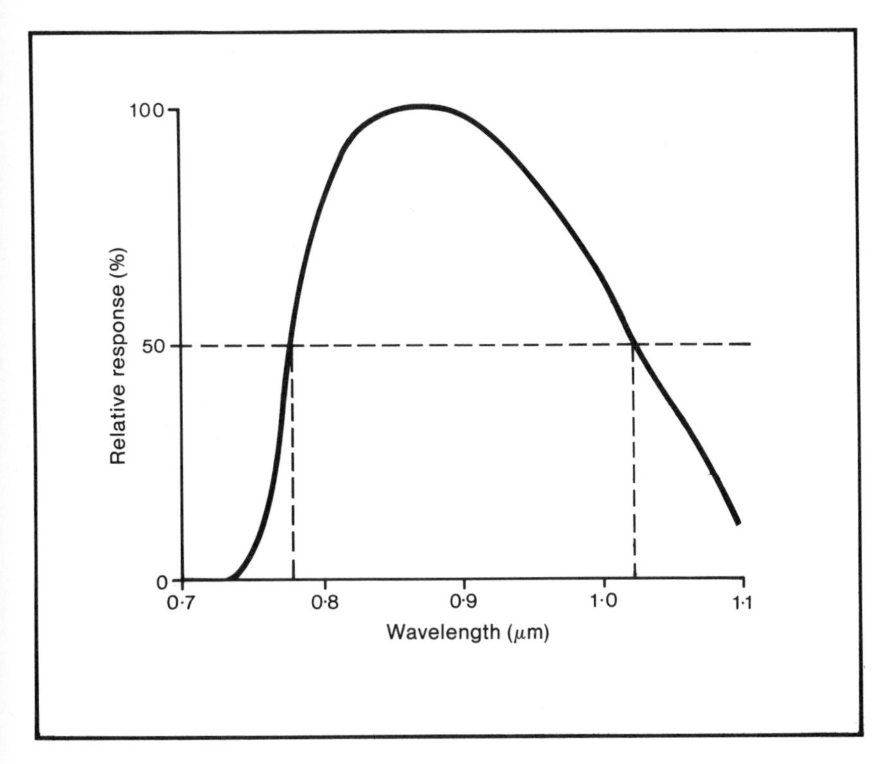


4.3 The spectral response curve for a near infrared filter/detector combination in a Milton Multiband Radiometer. The points of 50 per cent relative response on the curve are 0.78μm and 1.02μm. (Data courtesy E.J. Milton)

qualitative rather than a quantitative manner. Duggin and Philipson (1985) have identified some of the problems when attempting a quantitative comparison of ground radiometer and satellite data, and these are listed below.

(1) Instrument spectral response. The shapes of the response curves of the two instruments are different.
(2) Instrument spatial response. The instantaneous field of view of the two instruments is different. The ground radiometer is typically viewing a ground pixel of c. 1m diameter.
(3) Atmospheric attenuation caused by atmospheric absorption.
(4) Atmospheric forward scattering which scatters energy from neighbouring pixels into the path of reflected energy travelling from a pixel to the satellite sensor.

## APPLICATION

Both qualitative and quantitative ground data collection procedures have to be embedded in a broader remote sensing strategy. A number of factors have to be considered including sampling design, field measurement, data tabulation and processing, and comparison with satellite data, and these issues are discussed below.

### *Sampling*

The location of sample points at which to collect ground data should be governed by a sampling strategy. A stratified random sample should be taken where possible, stratified either spatially or by terrain cover type. Figure 4.4 shows an example of a stratified random sampling scheme employed for ground data collection on Glaisdale Moor in the North York Moors. Twenty-five sample sites were chosen randomly for each of three soil types in a study designed to evaluate the utility of satellite remote

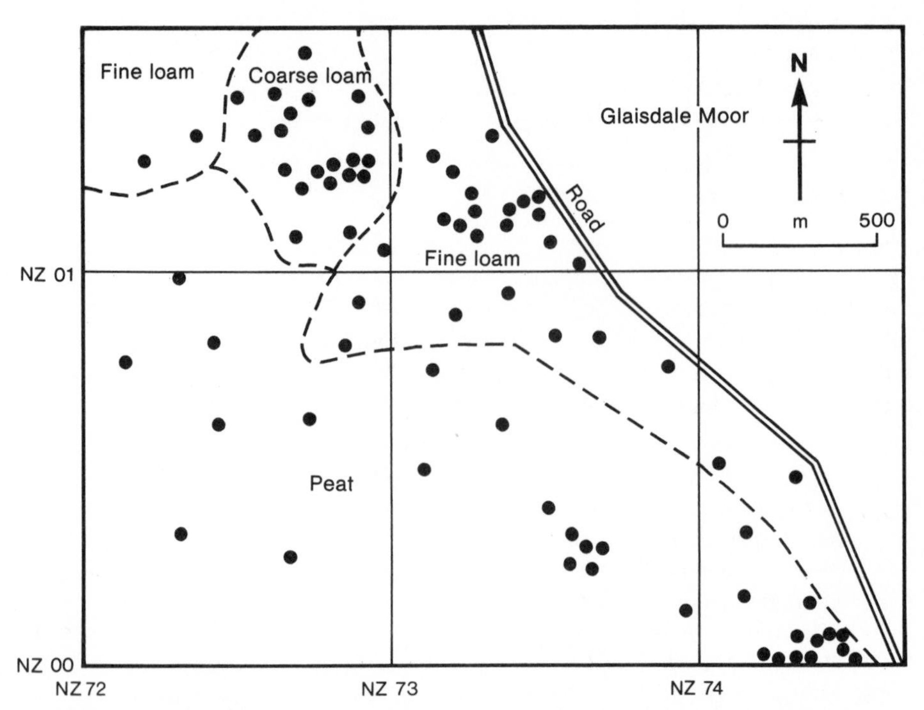


4.4 A stratified random sampling scheme employed for ground data collection on Glaisdale Moor, North York Moors. (Courtesy M.S. Alam)

sensing in the analysis of soil erosion. A random sampling scheme may not always be practicable in the field for reasons of access, but field measurements solely at accessible roadside locations should be avoided for two reasons:

(1) The terrain type may be unrepresentative because of the presence of the road.
(2) In later, quantitative analysis classification models often assume data collected on a random basis.

Justice and Townshend (1981) discuss the design of sampling schemes further.

### *Field measurement*

At the sample sites identified a ground radiometer will be set up as shown in Figure 4.1. This will allow measurement of reflectance from an area below the radiometer sensor. The area will depend upon the angular view of the radiometer and the height of the sensor, but for a Milton Multiband Radiometer at 2m above the ground surface this area is typically a circle of c. 1m diameter. Even an area as small as this will have small-scale terrain cover variations within it, so ancillary qualitative information about the terrain cover characteristics must also be collected (see Table 4.1).

### *Data tabulation and processing*

The field data will need to be recorded in a systematic form. Table 4.2 shows an example of a data table for a ground radiometer equipped with four channels. From these data the bidirectional reflectance for each of the four channels can be calculated using equation 4.1. The bidirectional reflectance data can then be plotted on scatter plots, often concentrating on one band taken from each of the visible and near infrared as shown in Figure 4.2.

### *Comparison with satellite data*

The ground radiometer data collected can be compared with satellite data at known points. This is difficult to achieve quantitatively because the satellite radiance data will include contributions from atmospheric effects and from neighbouring pixels, and also will be collected for a pixel much larger than the ground radiometer pixel (see above). A measure of comparison can be achieved by analysing the satellite radiance data for the known ground points and comparing the analysis with that performed for the ground reflectance data. Alternatively the ground data can be averaged over a large number of sites of similar terrain cover in order to compare with satellite data also averaged over large areas. This is particularly useful when dealing with meteorological satellite data of the Earth's surface.

## Longer wavelengths

The discussion in this chapter has so far concentrated upon ground data collection at visible and near infrared wavelengths. Thermal infrared and microwave remote sensing also require ground data for comparison with the satellite remote sensing data, but commonly the requirements are more difficult to achieve than for the shorter wavelengths.

### THERMAL INFRARED

Temperature data collected either *in situ* with ground thermometers or by using a thermal radiometer can serve as ground data. However, with thermal infrared remote sensing a number of other properties are important and these include soil moisture, vegetation moisture, evapotranspiration rates and airflow. The FIFE experiment in the USA gives a useful illustration of the types of ground data collection undertaken in thermal infrared remote sensing.

FIFE is the First ISLSCP Field Experiment taking place in 1986–7 (Ohring and Sellers 1984) and ISLSCP is the International Satellite Land Surface Climatology Project. The objective of FIFE is the development and validation of methods to convert satellite-observed radiances to climatological information, and to achieve this the emphasis is on the surface energy balance and on thermal infrared remote sensing. An extensive ground data collection effort is taking place in the US Mid-West, and a list of some of the variables measured at the ground surface is given in Table 4.3.

### RADAR

The equivalent to the radiometer for radar remote sensing is the scatterometer. This is an instrument mounted on a platform which emits pulses of microwave energy from a dish antenna directed at the surface. The scatterometer emits energy in a known waveband (say C-band) and can be

*TABLE 4.3: In situ measurement variables for the First ISLSCP Field Experiment (FIFE). The long-term measurements are taken over a one-year period and the short-term measurements during a set of two-week intensive experiments*

| *Long-term measurements* | *Short-term measurements* |
|---|---|
| Air temperature | Surface albedo |
| Atmospheric humidity | Solar insolation |
| Wind speed and direction | Net shortwave radiation |
| Precipitation | Eddy correlation |
| Soil temperature | Bidirectional reflectance |
| Soil moisture | Surface emissivity |
| Net radiation | Soil moisture profiles |
| Incoming and reflected solar insolation | Soil heat flux |
| Long wave radiation | Vegetation canopy temperature |
| | Organic matter |

*Source*: Ohring and Sellers (1984)

tilted at a range of incidence angles at the surface below. The return from the surface is the backscatter and this is also measured by the microwave dish antenna. An illustration is shown in Figure 4.5.

Ground scatterometers have been employed in the USA, Holland, France, West Germany and Japan for understanding the mechanisms of radar scattering by ground surfaces such as crops and bare soils (Zoughi *et al.* 1985; Wu *et al.* 1985; Kobayashi and Hirosawa 1985).

One form of ground information used in radar remote sensing but not normally in visible and near infrared remote sensing is the use of ground calibration points. Because radar provides active illumination of the surface, then targets of known characteristics can be located on the ground in the imaging area of the satellite radar, and these targets will return the radar signal to the satellite in a known way. The simplest target is a corner reflector which returns a maximum of energy to the satellite. On the radar images these corner reflectors appear as bright spots (Barber 1985). A second type of reflector is a Luneberg lens which consists of concentric spherical shells and a reflecting surface on its rear side. The concentric shells focus the incoming radar energy on to the rear side of the Luneberg lens and return it in the direction of its origin. Because the corner reflector and the Luneberg lens have known characteristics then the backscatter at points in an image can be calibrated. In addition, the ground reflectors also easily identify known ground points by their bright signal and so are useful for geometric rectification purposes.

During the SIR-B experiment in 1984 radar targets were buried below the surface in Nevada to investigate the depth of penetration of the L-band

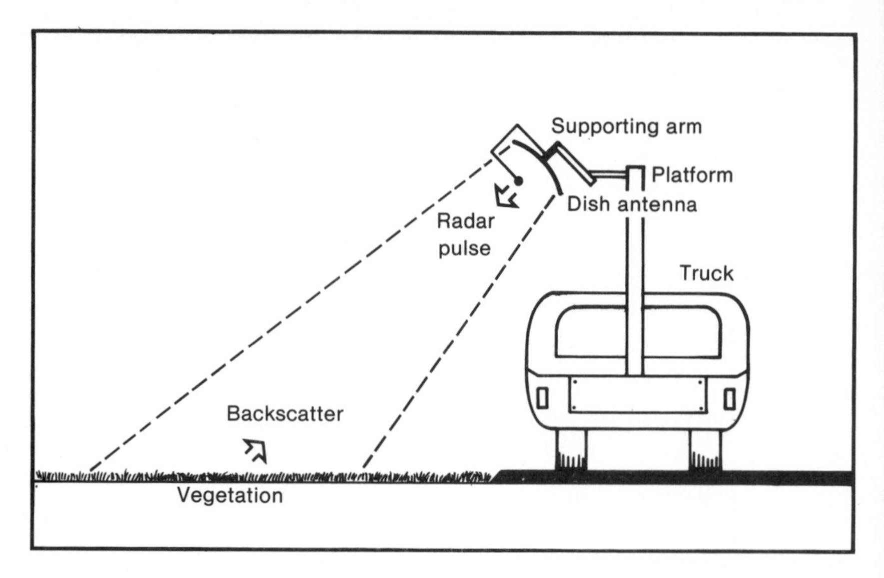


4.5 Illustration of a ground-based scatterometer. The microwave dish antenna emits a pulse of radar energy which is backscattered by the ground surface and returned to the antenna

signals. The results of this target detection experiment indicated that there was radar penetration of c. 1m depth even in semi-arid scrub conditions.

## Airborne data collection

Collection of complementary remote sensing data can be achieved using aircraft. Airborne systems can provide high-resolution imagery of the ground so that detailed maps of the terrain cover can be prepared for comparison with satellite remote sensing. The spatial resolution of airborne systems is high, with pixel sizes often less than 10m. As an illustration the Natural Environment Research Council has flown a series of airborne campaigns over selected sites in the UK with a Daedalus scanner (Williams 1984). The scanner has eleven channels and allows the simulation of a number of satellite sensor systems, including Landsat Thematic Mapper and SPOT HRV, but at a much higher spatial resolution. The eleven channels of the Daedalus scanner are listed in Table 4.4, and Figure 4.6 shows an example of the airborne scanner imagery for part of Morton Fen in eastern England.

*TABLE 4.4: The eleven channels of the Daedalus 1260 Airborne Thematic Mapper scanning radiometer, together with the approximate satellite equivalents*

| | *Daedalus channel* | *Wavelength range (μm)* | *Approximate satellite equivalents* | | | |
|---|---|---|---|---|---|---|
| | | | *Landsat TM* | *Landsat MSS* | *SPOT HRV* | *Noaa AVHRR* |
| | 1 | 0.42 – 0.45 | | | | |
| | 2 | 0.45 – 0.52 | TM1 | MSS4 | XS1 | |
| Visible | 3 | 0.52 – 0.605 | TM2 | | | |
| | 4 | 0.605 – 0.625 | | MSS5 | XS2 | AVHRR1 |
| | 5 | 0.63 – 0.69 | TM3 | | | |
| | 6 | 0.695 – 0.75 | | MSS6 | | |
| Near infrared | 7 | 0.76 – 0.90 | TM4 | MSS7 | XS3 | AVHRR2 |
| | 8 | 0.91 – 1.05 | | | | |
| | 9 | 1.55 – 1.75 | TM5 | | | |
| Middle infrared | 10 | 2.08 – 2.35 | TM7 | | | |
| Thermal infrared | 11 | 8.5 – 13.0 | TM6 | | | AVHRR4+5 |

*Source*: Williams (1984)

4.6 Airborne scanner image of part of Morton Fen in eastern England taken on 1 May 1986 using a Daedalus ATM 1260 scanning radiometer in band 11 (8.5–13.0μm). The pixel size is c. 10m. The road and field patterns are visible, as are crop marks which underlie some of the central fields. (Courtesy Natural Environment Research Council)

Before the SPOT satellite was launched simulation campaigns were held to collect airborne data to simulate the satellite data. Examples of the use of SPOT simulation data in the UK and the USA can be found in Borengasser and Taranik (1985), Buchan and Hubbard (1986), De Gloria (1985) and Essery and Wilcock (1986).

The use of airborne data provides a useful link between the *in situ* ground data and the satellite data, and can aid both the qualitative and quantitative interpretation of the satellite remote sensing data.

*Chapter 5*

# Satellite systems

## Introduction

A large number of satellites for remote sensing have been launched since 1960. Because many of the earlier satellites have been significantly superseded in their capabilities this chapter concentrates on recent and current satellite systems. Table 5.1 lists the main remote sensing systems currently or recently flown in space, and readers are recommended to refer to Table 5.1 in relation to each section of this chapter. The theme of the description in Table 5.1 is followed in this chapter. That is, the progression is from low spatial resolution systems, such as the geostationary weather satellites, through to the highest spatial resolution systems expressed in the Landsat Thematic Mapper, the SPOT HRV and the Large Format Camera. The corollary of this progression from low to high in spatial resolution is a reverse progression from high to low in temporal resolution. The geostationary weather satellites with a low spatial resolution (c. 5km pixels) commonly produce two images per hour of the same area, while the Landsat Thematic Mapper can only produce at a maximum one image of the same area once every sixteen days. This trade-off between spatial and temporal resolution is an important characteristic of satellite remote sensing and has a significant effect on the applications which are suitable for the different data sources.

Two principal orbit types are available to the designers of satellite remote sensing missions: polar orbits or geostationary orbit. The polar orbit is normally a low Earth orbit, sometimes termed LEO, where satellites fly at c. 200–1000km altitude viewing a swath of the Earth's surface below. The orbit takes the satellite over or near the north and south poles and the orbit is in the plane of the sun or sun-synchronous: that is, the orbit remains in a constant plane relative to the sun while the Earth spins below. On each

*TABLE 5.1: Principal recent and current satellite remote sensing systems*

| *Satellite*<br>*Instrument(s)*<br>*Country* | *Launch date(s)* | *Image repeat possibility* | *Wavelength(s)/ frequency* | *Spatial resolution* | *Swath width* | *Main applicatio* |
|---|---|---|---|---|---|---|
| GOES series Visible and Infrared Spin-Scan Radiometer (VISSR) USA | Series from October 1975 | 19 minutes | Visible<br><br>Thermal IR | 0.8km<br><br>6.9km | Full Earth disc or quarter disc | Meteorolog |
| Meteosat series Radiometer<br>Meteosat 1<br>Meteosat 2<br>ESA | <br><br>November 1977<br>June 1981 | 30 minutes | Visible<br>Middle IR<br>Thermal IR | 2.4km<br>5km<br>5km | Full Earth disc or sectors | Meteorolog<br>Vegetation<br>Environmer<br>hazards |
| GMS series (Himawari) Visible and Infrared Spin-Scan Radiometer Japan | Series from July 1977 | 30 minutes | Visible<br>Thermal IR | 1.25 or 4km<br>5 or 7km | Full Earth disc or quarter disc | Meteorolog |
| Insat I series Very High Resolution Radiometer India | IA April 1982 (abandoned September 1982)<br>IB August 1983 | 30 minutes | Visible<br><br>Thermal IR | 2.75km<br><br>11km | Full Earth disc | Meteorolog |
| Tiros-N/Noaa series Advanced Very High Resolution Radiometer (AVHRR) USA | Series from October 1978 | 12 hours | Visible<br>Near IR<br>Middle IR<br>Two thermal IR bands | 1.1km | 2,400km | Meteorolog<br>Oceanograph<br>Hydrology<br>Vegetation |
| Tiros-N/Noaa series Tiros Operational Vertical Sounder (TOVS) USA | Series from October 1978 | 12 hours | Twenty IR bands<br>Three IR bands<br>Four micro-wave bands | 17.4km<br>147.3km<br>105km (at nadir) | | Meteorolog |
| Defense Meteoro-logical Satellite Program (DMSP) Operational Linescan System (OLS) USA | Block 5D series from September 1976 | 12 hours | Visible and thermal IR | 0.6km | | Meteorolog |

| *ellite<br>trument(s)<br>untry* | *Launch date(s)* | *Image repeat possibility* | *Wavelength(s)/ frequency* | *Spatial resolution* | *Swath width* | *Main applications* |
|---|---|---|---|---|---|---|
| eteor series<br>SR | Series from<br>1969 | 12 hours | Visible<br>Thermal<br>IR | | | Meteorology |
| smos series<br>ıltispectral scanner<br>Kosmos 1500<br>Kosmos 1602<br>SR | <br><br>September 1983<br>September 1984 | 3 days | Four bands<br>visible and<br>near IR | 1.5km | 1930km | Meteorology<br>Land<br>applications |
| ndsat series<br>A<br>ultispectral scanner<br>ISS)<br>Landsat 1<br>Landsat 2<br>Landsat 3<br>Landsat 4<br>Landsat 5 | <br><br><br><br>July 1972<br>January 1975<br>March 1978<br>July 1982<br>March 1984 | <br><br><br><br>18 days<br>18 days<br>18 days<br>16 days<br>16 days | <br><br>Four bands in<br>visible and<br>near IR<br>Thermal IR<br>(Landsat 3<br>only) | <br><br>79m<br><br><br>237m | <br><br>185km | <br><br>Land use<br>Vegetation<br>Geology<br>Geomor-<br>phology<br>Hydrology |
| Landsat 4<br>Landsat 5<br>ematic Mapper<br>M) | July 1982<br>March 1984 | 16 days<br>16 days | Six bands<br>visible, near and<br>middle IR<br>Thermal IR | <br>30m<br><br>120m | 185km | Land use<br>Vegetation<br>Geology<br>Geomor-<br>phology<br>Hydrology<br>Cartography |
| OT<br>gh Resolution<br>sible (HRV)<br>ance | February 1986 | 2.5 days (off-<br>nadir) | Visible<br>Near IR<br>Panchromatic | 20m<br><br>10m | 60km | Land use<br>Agriculture<br>Cartography<br>Exploration |
| asat<br>nthetic Aperture<br>dar (SAR)<br>A | June 1978 | Limited cover | 23.5cm (L-<br>band) | 25m | 100km | Oceanography |
| ace Shuttle<br>uttle Imaging Radar<br>SIR-A<br>SIR-B<br>A | <br><br>November 1981<br>October 1984 | Limited cover | 23cm (L-<br>band) | <br><br>40m<br>30m | <br><br>50km<br>20–50km | Geology<br>Geomor-<br>phology<br>Soils<br>Land use<br>Oceanography |

| *Satellite Instrument(s) Country* | *Launch date(s)* | *Image repeat possibility* | *Wavelength(s)/ frequency* | *Spatial resolution* | *Swath width* | *Main applications* |
|---|---|---|---|---|---|---|
| Space Shuttle<br>Metric Camera<br>ESA | November 1983 | Limited cover | Panchromatic | c.20m | 190km<br>Stereo | Topographic mapping (1:50,000–1:100,000 scales) |
| Space Shuttle<br>Large Format Camera<br>USA | 1984 | Limited cover | Panchromatic<br>Colour<br>Near IR | c.20m | | Cartography (1:50,000 scale) |
| Space Shuttle<br>MOMS<br>West Germany | June 1983<br>February 1984 | Limited cover | Visible<br>Near IR<br>Thermal IR | 10–20m | 140km<br>Stereo | Vegetation<br>Land use<br>Hydrology<br>Geology |

*Sources*: Allison and Schnapf (1983), National Remote Sensing Centre (1985), Curran (1985), *Flight International* 12 January 1985

satellite orbit the Earth has rotated below the spacecraft (commonly c. 30° of longitude) so a coverage of the globe can be obtained by successive orbits.

The geostationary orbit, or GEO, is a high orbit where satellites are located above the equator at a distance of 35,900km, which is the same as that used by the communications satellites. At this distance the orbital speed of the satellite is at the same speed as the Earth's rotation (15° of longitude per hour) and so the satellite appears to be fixed above the same point on the surface of the Earth.

Figure 5.1 shows a combination of the polar and geostationary orbits. Thus far the orbits for all land and ocean remote sensing satellites have been polar orbits, while the weather satellites have used either polar or geostationary orbits.

## Geostationary weather satellites

The geostationary orbit is the principal orbit for communications satellites, and so in December 1966 NASA took advantage of the launch of a communications satellite, the first Applications Technology Satellite (ATS 1), to orbit a spin-scan camera to demonstrate the capability of providing a

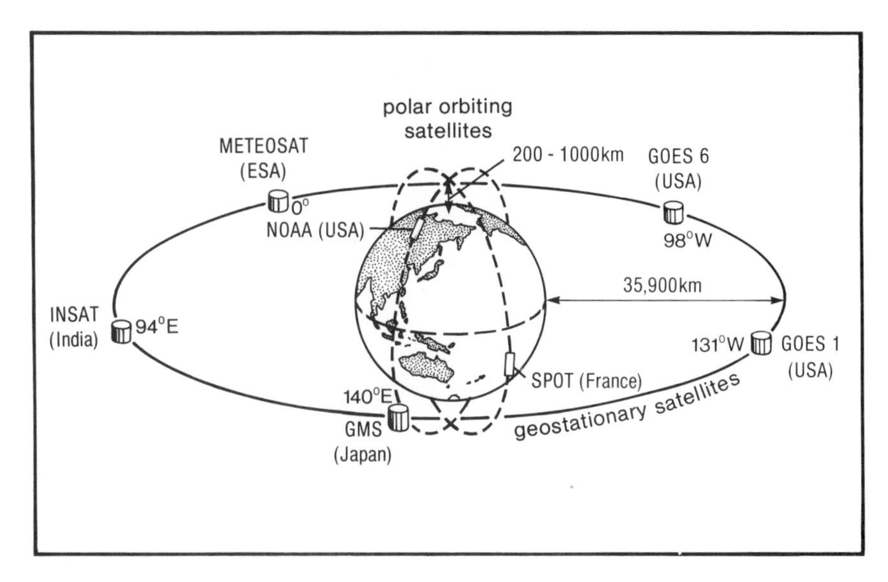


5.1 Polar and geostationary orbits used for meteorological and Earth observation satellites. (Modified from ESA 1985)

picture of a disc of the Earth every twenty minutes (Allison and Schnapf 1983). A further demonstration was flown on ATS 3 in November 1967, and this experience was translated into an operational programme with the launch of the Synchronous Meteorological Satellite (SMS 1) in May 1974. SMS 1 carried a Visible and Infrared Spin-Scan Radiometer (VISSR) to record images of the Earth's disk in the visible (0.66–0.7μm) and thermal infrared (10.5–12.6μm) wavebands every nineteen minutes. The first of the operational SMS satellites was launched in October 1975 and the programme was renamed GOES, the Geostationary Operational Environmental Satellite, owned and operated by the US National Oceanic and Atmospheric Administration (NOAA). Since then NOAA has launched five further GOES satellites and has moved them to different locations on their equatorial orbits as demands have changed. The operational GOES system consists of three satellites, but the series has been plagued by sensor failures (*Flight International* 12 January 1985). Sensor failures hit GOES 4 in November 1982 and GOES 5 in July 1984, and for a time the only GOES satellites in operation were GOES 1, operating at visible wavelengths and so of use only during daylight, and GOES 6 which was fully operational. The basic payload of GOES is the VISSR, but after GOES 3 a microwave atmospheric sounder was added to the payload to provide vertical atmospheric profile information.

In Europe a geostationary weather satellite programme initiated by the

French took shape in the 1970s. In 1973 the European Space Agency (ESA) agreed to the development of two Meteosat satellites, and in November 1977 Meteosat 1 was placed into a geostationary orbit above 0° longitude 0° latitude, i.e. above a position in the Gulf of Guinea south of West Africa. Meteosat 1 failed in 1979 but was replaced in 1981 by Meteosat 2. The satellite carries a radiometer operating in three bands as shown in Table 5.2. The visible/near infrared and the thermal infrared wavebands are the standard wavebands, but Meteosat also carries a sensor with a 5.7–7.1μm wavelength to map variations in atmospheric water vapour. The imagery from Meteosat extends to c. 55° north and south of the equator, so it provides good coverage of Africa and southern Europe but poor coverage of northern Europe. An illustration of imagery from the three wavebands is shown in Figure 5.2, and such coverage is available from the satellite once every thirty minutes.

*TABLE 5.2: Meteosat wavebands*

| *Band* | *Wavelength range (μm)* | *Resolution (km) at 0° latitude 0° longitude* |
|---|---|---|
| Visible/near infrared | 0.4–1.1 | 2.5 |
| Water vapour | 5.7–7.1 | 5.0 |
| Thermal infrared | 10.5–12.5 | 5.0 |

Europe has established the EUMETSAT international meteorological organisation to operate an operational Meteosat programme of three satellites to be launched by Ariane rockets in 1988, 1989 and 1990.

Japan also saw the virtue of geostationary weather satellites and in July 1977 launched its own (GMS 1), also called Himawari 1, and its successor GMS 2 in August 1981. GMS 2 failed in January 1984 but was replaced in August 1984 by GMS 3. Each GMS satellite was positioned over the western Pacific at 140° E longitude, 0° latitude, and they have been particularly important in warning of storms headed for eastern Asia.

A recent participant in geostationary weather satellites is India. The Insat I series of geostationary satellites is unique in combining three functions: telecommunications relay, direct broadcast and weather imaging. The first Insat (IA) was launched in April 1982 and was equipped with a Very High Resolution Radiometer with a visible and a thermal infrared channel which had pixel sizes of 2.75km and 11km respectively. Insat IA was placed in a position above the equator at 74° E longitude, but failed in September 1982 because of exhaustion of its altitude control fuel. In August 1982 the replacement Insat IB was launched from NASA's Space Shuttle

(b)

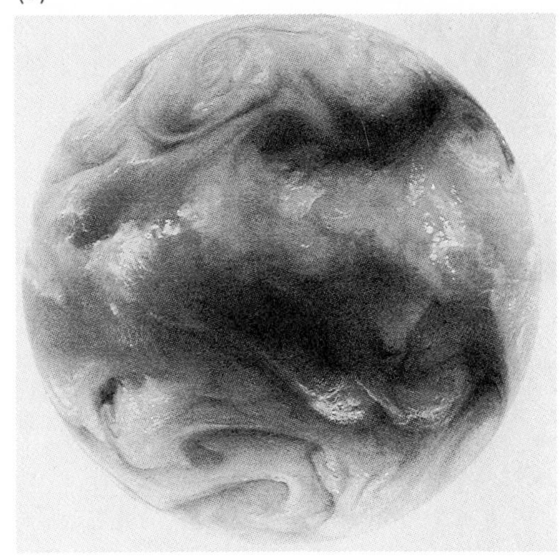

(c)

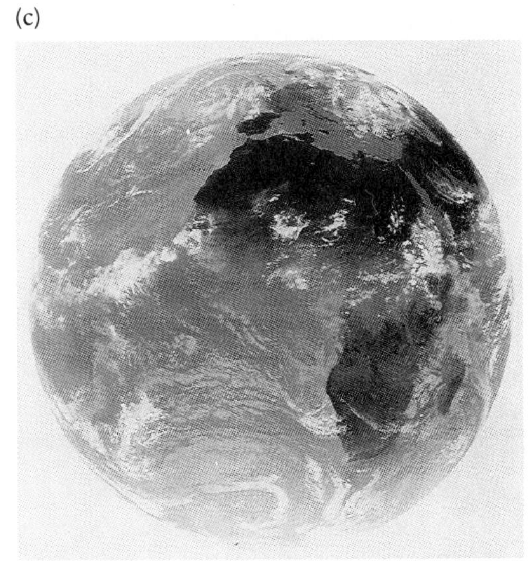

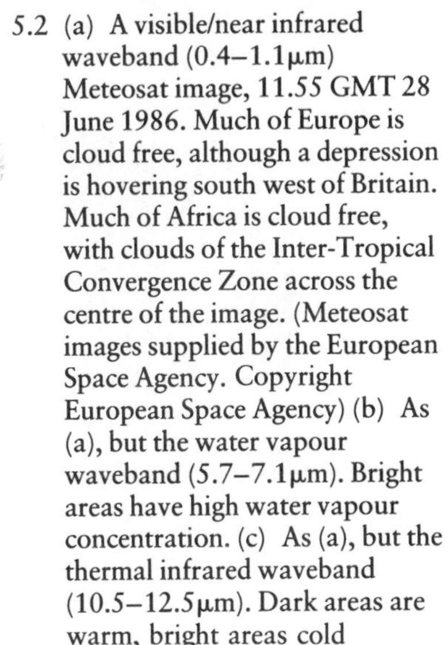

5.2 (a) A visible/near infrared waveband (0.4–1.1μm) Meteosat image, 11.55 GMT 28 June 1986. Much of Europe is cloud free, although a depression is hovering south west of Britain. Much of Africa is cloud free, with clouds of the Inter-Tropical Convergence Zone across the centre of the image. (Meteosat images supplied by the European Space Agency. Copyright European Space Agency) (b) As (a), but the water vapour waveband (5.7–7.1μm). Bright areas have high water vapour concentration. (c) As (a), but the thermal infrared waveband (10.5–12.5μm). Dark areas are warm, bright areas cold

and located in geostationary orbit at 94° E longitude. India plans to launch Insat IC in the near future to replace the failed IA.

The geostationary satellites of the USA, Europe and Japan were used in a joint global data collection exercise in the First GARP Global Experiment in 1978–9. The satellites involved are listed in Table 5.3. The GOES-IO satellite was moved from a position over the western hemisphere to take up a position agreed but not filled by the GOMS geostationary satellite of the USSR. The position has since been filled permanently by the Indian satellite Insat. As each geostationary satellite views approximately one-third of the Earth then the five satellites with overlapping image areas provided a

*TABLE 5.3: The geostationary weather satellites which were used in the First GARP Global Experiment (FGGE) in 1978–9*

| *Satellite* | *Location* | *Longitude* |
|---|---|---|
| Meteosat (Europe) | West Africa | 0° |
| GOES-E (USA) | Western Atlantic | 70° W |
| GOES-W (USA) | Eastern Pacific | 140° W |
| GMS (Japan) | Western Pacific | 140° E |
| GOES-IO (USA) | Indian Ocean | 70° E |

complete coverage of the Earth between latitudes 55° north and south, and did so with a high temporal frequency of c. two images per hour.

One important advantage which geostationary satellites offer is the ability to see change. With an image collected every thirty minutes then analyses can be prepared of cloud development, air movement as revealed by cloud motion, and sea surface temperature changes. Such change analysis is particularly important for monitoring the tracks and the development of tropical hurricanes (Smith *et al.* 1986). In 1985 thousands of lives were saved in Bangladesh by meteorologists being able to follow and forecast the development of an intense Bay of Bengal cyclone as it approached the Bangladesh coast. The forecast allowed the local population time to move northwards and inland to safety.

In September 1985 two hurricanes hit the USA. In early September Hurricane Elena hit the coast of Mississippi with winds in excess of 100mph. The ability to monitor from the GOES satellite the growth and movement of the hurricane over the Gulf of Mexico enabled the authorities to issue advance warnings in several US southern coastal states and for over one million people to be evacuated from their homes before the hurricane struck. In late September Hurricane Gloria, one of the most severe Atlantic storms this century, seriously affected the eastern seaboard of the USA. Its progress was plotted from GOES imagery as it travelled northwards from North Carolina to New England, generating winds of up to 130mph. In Connecticut and Massachusetts a state of emergency was declared, 200,000 people were evacuated from Long Island, and along the eastern seaboard of the USA more than a quarter of a million people were moved inland for safety. The effects of Hurricane Gloria were particularly severe in New York City where the stock exchanges, the World Trade Center, banks, offices, shops and airports were closed and people evacuated before the hurricane struck.

# Polar orbiting weather satellites

## FIRST GENERATION

The initial emphasis in satellite remote sensing was on the atmosphere, not on the Earth's surface. The USA launched its Tiros 1 (Television and Infrared Observation Satellite) on 1 April 1960, and scientists quickly appreciated the value of a large-area view of cloud systems, particularly over remote ocean areas. More Tiros satellites followed the short-lived Tiros 1 experiment, and between 1960 and 1965 a total of ten Tiros satellites were launched, each equipped with a small television camera, and in some missions a thermal infrared scanning radiometer and an Earth radiation budget experiment. The main limitation of the Tiros satellites was the extent of their coverage. The orbit of the Tiros 1–4 satellites was inclined at 48° to the equator, and of the Tiros 5–8 satellites 58° (Barrett 1974). This meant that the satellites orbited only as far as latitudes 48° and 58° respectively, so that the data coverage was good in tropical and sub-tropical regions, but the northern mid-latitudes and the polar regions were neglected.

The Tiros series of essentially experimental satellites was followed in 1966 by an operational series of satellites termed the Tiros Operational System, which was designed to provide routine daily observations on a worldwide basis without interruption of the data supply. The satellites themselves were termed Essa satellites after their operators, the US Environmental Science Services Administration. Nine Essa satellites were launched between 1966 and 1969 and provided a good source of imagery on a global basis. Global data collection was achieved because the Essa satellites orbited the Earth at an angle of c. 100° to the equator so that on each orbit the satellite passed close to the poles (within 10° of latitude). Mosaics of images were made on a polar stereographic projection on a routine basis for the northern and southern hemispheres centred on their respective poles, and although this did not provide a simultaneous view of hemispheric weather systems it did provide an opportunity to see the overall spatial structure of the atmosphere as revealed by the cloud patterns.

The even-numbered Essa satellites (Essa 2, 4, 6, 8) provided direct transmission of their image data to ground receiving stations around the world by Automatic Picture Transmission (APT). This allowed many users to build low-cost receiving stations to receive weather satellite pictures of their local area. The extent of the reception was constrained by the satellite being in line of sight of the receiving station. The odd-numbered Essa satellites (Essa 1, 3, 5, 7, 9) stored their data on tape recorders carried on board the satellites. The stored data were later transmitted to the ground when the satellite flew over one of the two receiving stations in Virginia and Alaska. This storage of data allowed weather satellite data for all parts of the globe to be collected in and disseminated from the USA.

## SECOND GENERATION

The second decade of weather satellite operation also saw the introduction of the second generation of meteorological satellites, the Improved Tiros Operational System (ITOS), the first of which was launched in 1970. Each ITOS satellite had the combined advantages of direct APT transmission and on-board data storage, and they could also collect imagery during the day (visible and thermal infrared) and the night (thermal infrared only) using scanning radiometers of medium (4km) and high (0.9km) spatial resolution. The ITOS satellites were termed Noaa on successful launch, named after their operators the National Oceanic and Atmospheric Administration, which was the successor to ESSA.

## THIRD GENERATION

The third generation of polar orbiting weather satellites and the current operational US system is the Tiros-N series. The first Tiros-N satellite was placed into orbit in 1978, and its successors became Noaa 6 and 7, followed by an Advanced Tiros-N satellite flown as Noaa 8, 9 and 10. The Tiros-N/Noaa satellite is equipped with an Advanced Very High Resolution Radiometer (AVHRR) which is a scanning radiometer with channels in the visible, near, middle and thermal infrared and a spatial resolution of 1.1km. The channels are shown in Table 5.4. The broad-band visible and near infrared channels of previous sensors have been split into separate visible (channel 1) and near infrared (channel 2) bands, and in the case of the five-channel instrument the thermal infrared has also been divided. The visible and near infrared channels are used to identify clouds, land–water boundaries, and snow and ice extent, and when the two channels are compared then an indication of ice and snow melting can be gained.

One advantage of the division of the visible and near infrared has been in land surface analysis. A comparison of the radiances in the two channels can be used as a measure of vegetation, and, given the daily global coverage of the Noaa satellites surface, vegetation and/or biomass can be monitored in cloud-free regions. Justice *et al.* (1985) give an analysis of global vegetation characteristics using AVHRR data, Tucker *et al.* (1985) analyse the

*TABLE 5.4: Wavebands of the Tiros-N/Noaa Advanced Very High Resolution Radiometer. Spatial resolution is 1.1km*

| *Channel* | *Four-channel instrument* | *Five-channel instrument* |
|---|---|---|
| 1 | 0.55 – 0.68μm | 0.58 – 0.68μm |
| 2 | 0.725– 1.10μm | 0.725– 1.10μm |
| 3 | 3.55 – 3.93μm | 03.55 – 3.93μm |
| 4 | 10.5 –11.5μm | 10.3 –11.3μm |
| 5 | | 11.5 –12.5μm |

*Sources*: Townshend and Tucker (1981), Harris (1985a)

seasonal vegetation changes in the African continent, and Nelson and Holben (1986) use the AVHRR data to identify deforestation in Brazil. NOAA produces a weekly composite map of a vegetation index for both hemispheres calculated from the visible and near infrared Noaa satellite data (Yates *et al.* 1986).

Data from the thermal infrared channels of Tiros-N/Noaa satellites are used to measure the temperature of the surface or the clouds beneath the satellite, and a combination of the two thermal infrared channels (channels 4 and 5 – see Table 5.4) can be used to calculate sea surface temperature (see chapter 10). The channel three waveband (3.55–3.93μm) is sensitive to atmospheric water vapour distributions and to high temperatures (>50°C) such as those created by surface fires (Muirhead and Cracknell 1984, 1985).

Data from the Noaa AVHRR sensors are available in three forms, and these are described below.

### *High Resolution Picture Transmission (HRPT)*

HRPT data are at the full resolution of 1.1km and can be picked up by a suitable ground receiving station when the satellite is in direct line of sight. HRPT is the successor to Automatic Picture Transmission.

### *Local Area Coverage (LAC)*

On each orbit up to ten minutes of full resolution data are recorded on board the satellite and can be replayed when in line of sight of a US ground station (Townshend and Tucker 1981). The ten minutes of data correspond to a ground strip some 4,000km long.

### *Global Area Coverage (GAC)*

The full resolution data are sampled to reduce the data quantity and allow global coverage. The sampling framework is a block of three rows and five pixels per row of full resolution data. One GAC pixel represents this block of fifteen pixels by the calculation on board the satellite of the average of the first four pixels in the first row of this 3 × 5 pixel block: this average is used as the GAC pixel value. This is not a satisfactory method of representing the average of a 3 × 5 pixel block, but it is the form adopted operationally by NOAA.

Figures 5.3 and 5.4 show examples of Noaa AVHRR images for the area of the British Isles. Both images are for 1 June 1985 at 13.36 GMT and were received from the Noaa 9 satellite at the University of Dundee by direct HRPT transmission. Figure 5.3 is a visible waveband image and Figure 5.4 a thermal infrared image. They both show clear conditions over most of Britain, with scattered cloud over northern Scotland and a cold front north of the country. On Figure 5.4 the thin lines orientated north-west–south-east over Scotland are the cold contrails of jet aircraft. The parallel lines of cloud north of Scotland are a wave formation set up in the westerly airflow in the wake of the Faroe Islands.

## VERTICAL SOUNDING

This discussion of weather satellites has so far concentrated on imaging sensors. These sensors are complemented on Noaa satellites by sensors designed to obtain information about the vertical structure of the atmosphere. The sensors operate at low spatial resolution (of the order of tens of kilometres) and are not designed as imaging systems. Noaa satellites now carry the Tiros Operational Vertical Sounder (TOVS) package which consists of three separate instruments:

High Resolution Infrared Radiation Sounder (HIRS2)
Stratospheric Sounding Unit (SSU)
Microwave Sounding Unit (MSU)

The TOVS data allow the calculation of atmospheric temperature profiles within the field of view from the surface to c. 30km altitude, water vapour content at three levels in the atmosphere and total atmospheric ozone content. Other sensors on the current Noaa satellites include an Earth radiation budget sensor, an instrument to measure the ultraviolet spectrum of the sun or of solar radiation reflected from the Earth, a data collection

5.3 *Left* A visible wavelength Noaa 9 AVHRR image of the British Isles, 13.36 GMT 1 June 1985. (Courtesy University of Dundee)
5.4 *Right* As Figure 5.3 but thermal infrared. Dark tones are warm, light tones are cool. (Courtesy University of Dundee)

system for interrogation of remote sea buoys, and a Sarsat search and rescue mission for assisting with the location of ships and aircraft in distress.

## OTHER WEATHER SATELLITES

The series of Tiros–Essa–Itos–Noaa satellites have all been operated by the US civilian sector, currently by NOAA, a branch of the US Department of Commerce. In parallel with these satellites has been the Defense Meteorological Satellite Program (DMSP) operated by the US military. The DMSP satellites routinely transmit data to the US Air Force and Navy ground terminals and to Navy carriers. The first DMSP satellite was launched in 1966 as the Block IVA model, and this was followed by twenty-three other satellites of six other models, taking us up to the current Block 5D-2. There are two wavebands and two operating modes on the DMSP Block 5D satellites. The two wavebands are visible and near infrared (0.4–1.1μm) and thermal infrared (10.5–12.6μm), and the two operating modes are fine resolution (0.6km pixels) and smooth resolution (2.8km pixels). The DMSP satellites provide the highest resolution meteorological satellite data available, although to civilian users only in image and not in digital form. The most recent Block 5D-2 F9 satellite carries in addition to the imaging system a Special Sensor Microwave/Imager (SSM/I) which is designed to measure ocean surface wind speed, ice coverage and age, cloud water content, rainfall and atmospheric vertical temperature and humidity profiles.

Alongside the operational satellites operated by the US civilian and military sectors there has also been a series of Nimbus satellites which have acted as platforms for experimental sensor packages. Seven Nimbus spacecraft were successfully launched between 1964 and 1978, and the last one (Nimbus 7) was only retired in 1984. The sensors carried on Nimbus 7 measured the microwave emissivity of clouds, atmospheric ozone, vertical gas and aerosol concentration, and coastal sea composition.

## Heat Capacity Mapping Mission

The first of the specifically land applications satellites discussed in this chapter is the Heat Capacity Mapping Mission (HCMM) launched in April 1978. The satellite was placed in a polar orbit (97.6 °) with equator crossing times of approximately 2am and 2pm (local time). This was designed to achieve the greatest temperature contrast in the surface and thereby measure the thermal inertia of the land surface (see chapter 8). The sensing instrument was the Heat Capacity Mapping Radiometer which had two channels: visible and near infrared (0.5–1.1μm) and thermal infrared (10.5–12.5μm). The visible channel had a pixel size of 0.5km × 0.5km and the thermal infrared channel 0.6km × 0.6km. The thermal inertia calculated from the night and day passes of the scanner were used extensively by geologists to discriminate rock types, and Elachi (1983) gives examples of dolomite and limestone discrimination in Death Valley, California using HCMM imagery. HCMM data were also used for geomorphic mapping, soil moisture and evapotranspiration measurement. In Europe analysis of HCMM data was largely performed within the TELLUS project which is described by Galli de Paratesi and Reiniger (1983).

# Landsat

Landsat satellites have provided probably the most useful source of satellite remote sensing data for geographers. The aim of Landsat is to provide repetitive acquisition of high resolution multispectral data on a global basis, and because of this the data have been used by a wide variety of environmental scientists concerned with the Earth's land surface and water bodies. The development of the Landsat programme was in a sense a spin-off from the US lunar programme when certain scientists in NASA realised the significance of viewing the Earth from space, as well as reaching the moon. The spectacular pictures of the Earth taken by astronauts using hand-held cameras on the Mercury and Gemini spacecraft clearly illustrated the value, the beauty and the excitement of space imagery of the Earth.

In 1966 the planning programme of what was to become Landsat was started, and in 1969 the programme termed the Earth Resources Technology Satellite (ERTS) was initiated. This title was given to the first satellite in the series (ERTS 1), but the name ERTS was changed to Landsat retrospectively shortly before the launch of the second satellite in the series. In 1970 NASA invited worldwide participation in the ERTS programme and sponsored several hundred principal investigators to work on the new data in a wide variety of subject areas, including: agricultural production; water resources; rangeland and forestry management; urban and regional planning and geology and mineral resources. In retrospect the level of participation, the commitment and the enthusiasm generated by the space age was remarkably high for a satellite which had not then been launched and which had a design lifetime of only one year.

Landsat has had two generations, and fourteen years after the launch of the first satellite the programme is now approaching a level of maturity in its second generation. The first generation comprised Landsats 1, 2 and 3, and the second generation Landsats 4 and 5. Landsats 6 and 7 are planned for launch in the 1990s (NOAA 1986) (see chapter 11).

## FIRST GENERATION LANDSAT: 1, 2, 3

Landsat 1 was launched to an orbital altitude of 913km on 23 July 1972 from the Vandenburg Air Force Base in California on a McDonnell-Douglas Delta rocket. The launch dates and lifetimes of the first three Landsat satellites are given in Table 5.5.

The platform employed for the first generation Landsats was a modified Nimbus type and is illustrated in Figure 5.5. The platform consists of two

*TABLE 5.5: Operation periods of the Landsat 1–3 satellites*

| *Satellite* | *Launch date* | *End of operation* |
|---|---|---|
| Landsat 1 | 23 July 1972 | 6 January 1978 |
| Landsat 2 | 22 January 1975 | 27 July 1983* |
| Landsat 3 | 5 March 1978 | 7 September 1983** |

*Source*: NOAA (1986)

*Interruption in 1979–80
**Standby mode from 31 March 1983

principal parts: the lower sensor ring which carries the instruments and the upper solar array panels which collect solar energy to power those instruments. The remote sensing payload of Landsat 1 consisted of a Return Beam Vidicon (RBV) camera and a Multispectral Scanner (MSS), and as with most satellite systems this represented a compromise between conflicting views (Freden and Gordon 1983). The original NASA and US Geological Survey (USGS) studies had recommended frame-type picture imagery for which the RBV camera was appropriate, while the US Department of Agriculture (USDA) wanted good spectral information for which the MSS scanning radiometer was more suitable. So both were carried.

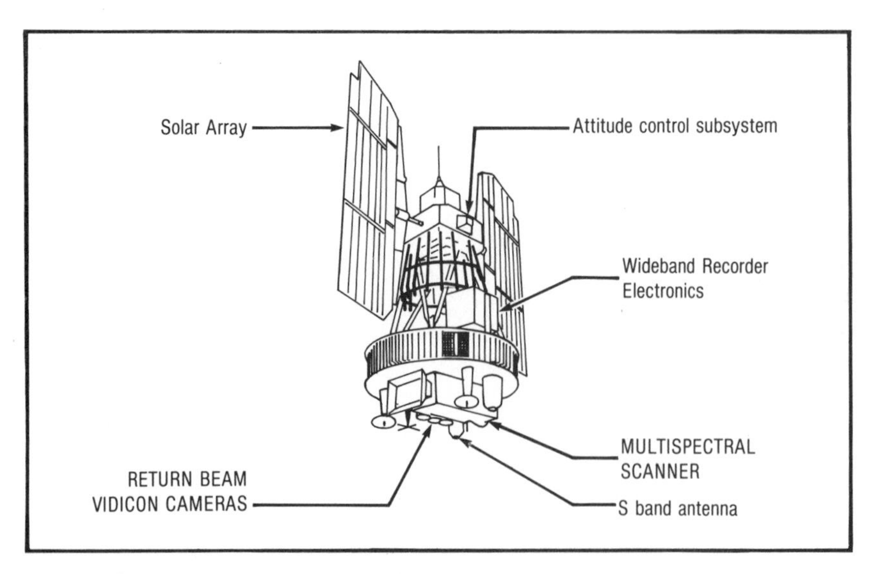


5.5 The first generation Landsat platform

The equator crossing time of the satellite was also a compromise. For geology and geomorphology (as represented by the USGS) an early morning or late afternoon equator crossing was useful because it gave a low sun angle and a good display of topography. For agricultural applications (as represented by the USDA) maximum illumination was needed and a noon pass was preferred. The compromise was an equatorial crossing time of about 9.30 am local time, which had the advantage of lower cloud cover in those areas prone to daily cloud build-up. For the northern mid-latitudes this gave a data acquisition time of around 10 am local time.

The near polar orbit (99°) and the altitude (913km) of the first generation Landsats gave an orbit pattern where each area on the Earth's surface was flown over once every eighteen days. This did not in practice mean that an image of each part of the globe was obtained once every eighteen days because of data storage, transmission and handling constraints, and because of high cloud cover in many parts of the tropics and the mid-latitudes. For many parts of Britain, for example, only one or two scenes per year have been collected, largely because of the high incidence of cloud cover.

*Multispectral Scanner (MSS)*

The Landsat Multispectral Scanner (MSS) is a scanning radiometer which scans 185km of the Earth's surface in a west–east direction across the ground track of the spacecraft. The diagram in Figure 3.2 in chapter 3 illustrates the Landsat MSS. The oscillating mirror of the scanner focuses the energy of six scan lines simultaneously onto an array of fibre-optic light pipes which each lead to an individual filter-detector. The array of fibre optics is 6 × 4 light pipes in size, corresponding to six scan lines and to the four wavebands of the sensor. The four wavebands of the MSS are listed in Table 5.6. The diameter of each optical fibre gives the physical control on the spatial resolution of the sensor. Each fibre end provides an angular instantaneous field of view (IFOV) of 0.986 milliradians, which from the 913km altitude of Landsat gives a ground IFOV or pixel size of 79m × 79m. Because of the way in which the signal for each detector is sampled there is an overlap between successive pixels. The centre of each pixel is 56m removed along a scan line from the centre of its predecessor, so any one 79m × 79m pixel area contains about 11.5m of the previous pixel and 11.5m of the next pixel.

An example of a Landsat 2 MSS band 7 image is shown in Figure 5.6. This is an image of northern England acquired on 27 May 1977. The image clearly shows the moorlands of the Pennines, the North York Moors and the Lake District, the conurbations of Tyneside and Teeside, the agriculture of the Vale of York, and part of the North Sea.

On Landsat 3 the MSS was extended to include a band 8 thermal infrared

*TABLE 5.6: Wavebands of the Landsat Multispectral Scanner*

| *Band* | *Wavelength range (μm)* | *Applications* |
|---|---|---|
| 4 | 0.5–0.6 | Sediment loads; shallow water |
| 5 | 0.6–0.7 | Vegetation; cultural features |
| 6 | 0.7–0.8 | Land/water separation |
| 7 | 0.8–1.1 (1.0* ) | Vegetation and geological studies |

*The upper limit of band 7 is normally quoted as 1.1μm, but has also been quoted as 1.0μm

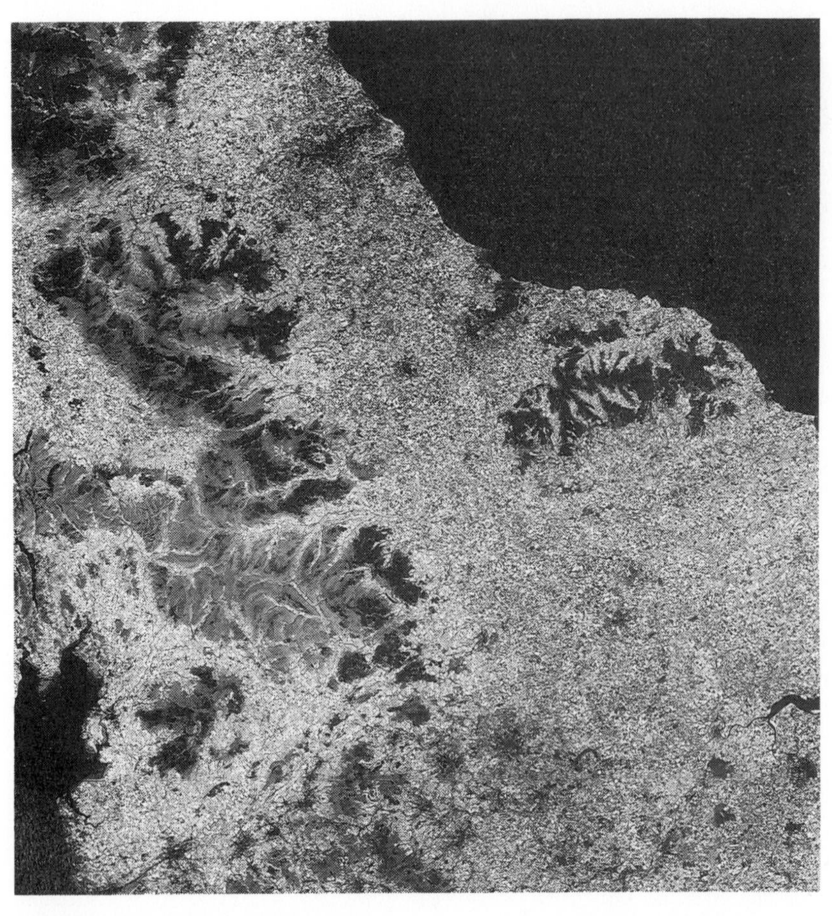

5.6 Landsat 2 MSS band 7 (0.8–1.1μm) image of northern England, 27 May 1977. (Courtesy RAE, Farnborough)

band of 10.4–12.5μm and a pixel size of 237m × 237m (equivalent to three 79m pixels of bands 4–7). Because of a limit to its thermal sensitivity and a lack of calibration few good scenes were collected using this band.

### *Return Beam Vidicon (RBV)*

The original Return Beam Vidicon (RBV) system flown on Landsats 1 and 2 was not a success. Three RBV cameras were mounted on the spacecraft, and all three were focused on the same ground area. The three cameras were sensitive to the following wavebands: 0.475–0.575μm, blue–green; 0.58–0.68μm, yellow–red and 0.69–0.83μm, red–near infrared. The band names were 1, 2, 3 respectively, which is why the MSS bands were named 4, 5, 6, 7. The spatial resolution of the early RBV was the same as the MSS (79m), although in practice this depended more on scene contrast than did the MSS, and the image frame area was also 185km × 185km.

However, this original RBV system was not a success and only forty-five hours of RBV operation were achieved with Landsats 1 and 2 compared to 7,462 hours of MSS operation. A different and more successful approach was taken with the RBV on Landsat 3. A panchromatic RBV with a broad spectral range of 0.505–0.750μm (green–near infrared) and an improved spatial resolution of 24m (maximum) was flown on Landsat 3. Two identical RBV cameras were mounted on the spacecraft, each viewing a ground area of 98km × 98km. Four overlapping scenes were registered to be approximately equivalent to a 185km × 185km MSS scene. The Landsat 3 RBV was much more successful than its predecessors, and operated for thousands of hours. Figure 3.1 in chapter 3 shows part of a Landsat 3 RBV scene of Kuwait and illustrates the high spatial resolution of the sensor: roads and coastal features are clearly visible, even from an altitude of 913km.

### *Data transmission*

When Landsat 1 was launched there were only four receiving stations: three in the USA and one in Canada. This meant that direct transmission of Landsat data could only take place for images of the North American region. To allow data to be collected from other parts of the globe the satellite carried a tape recorder which recorded the image data for later playback when the satellite passed over one of the US ground receiving stations. In the period since the launch of Landsat 1 a number of countries have built Landsat receiving stations: a map of these stations and their coverage areas is shown in Figure 5.7.

In the USA Landsat data are archived at the EROS Data Center, Sioux Falls, South Dakota. The receiving stations outside the USA have their own

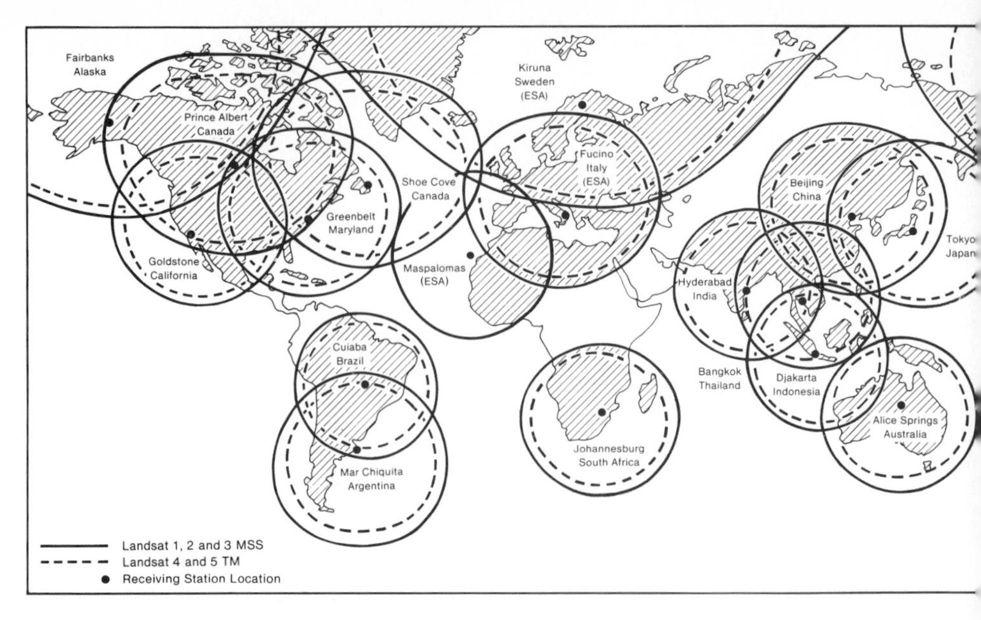


5.7 The Landsat receiving stations and their coverage areas

data archives, and in Europe this is the Earthnet archive which is accessible in Britain through the National Remote Sensing Centre, Farnborough. Appendix 2 gives the addresses of the Landsat data distribution centres.

## SECOND GENERATION LANDSAT: 4 AND 5

The second generation of the Landsat programme began with the launch of the fourth satellite in the series on 16 July 1982, and continued with the launch of Landsat 5 on 1 March 1984. Landsat 4 was a completely different design to the Nimbus-type platform of Landsats 1–3, and Figure 5.8 is a diagram of the satellite. For this second generation of Landsat a lower orbit of c. 700km was chosen so that it would be possible to recapture the satellite for repairs using the Space Shuttle, and a shorter revisit cycle was achieved. Landsats 4 and 5 each pass over the same area of the Earth once every sixteen days, but as with the first generation system the real image acquisitions are achieved with a far lower frequency than this.

Landsats 4 and 5 carry an MSS of the same design as the early Landsats, except (1) the ground pixel size is 82m × 82m, and (2) the band numbering is changed from 4, 5, 6, 7 to 1, 2, 3, 4. The RBV camera has been omitted and a new sensor introduced, the Thematic Mapper.

### *Thematic Mapper (TM)*

The Thematic Mapper (TM) is a high spatial resolution sensor with an improved radiometric sensitivity. The bands were chosen with the ultimate

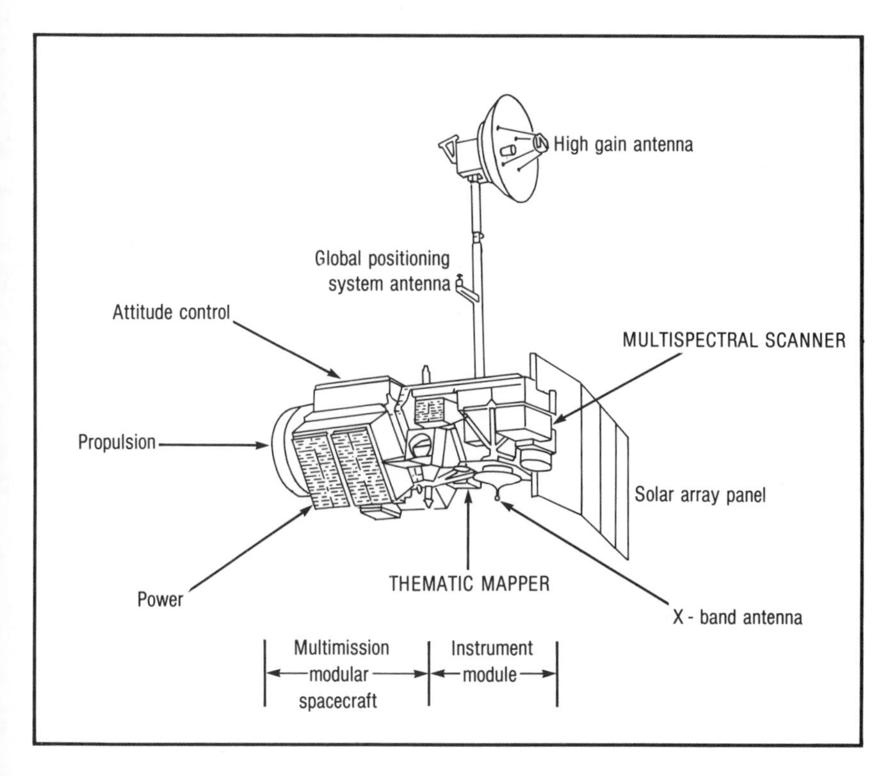


5.8 The second generation Landsat platform

applications of the data more firmly in mind, and in Table 5.7 are listed the spectral characteristics of the TM with the principal applications associated with each band. Information on the MSS is given for comparative purposes.

One key improvement in the Thematic Mapper is its spatial resolution. The pixel size has been reduced to 30m for six bands, and the thermal infrared band has 120m pixels. The band numbering is not in the sequence of longer wavelengths because the 2.08–2.35μm band was finally included only at a late stage after much debate and pressure from the geological community.

The TM is a scanning radiometer, but it does not employ fibre optics as does the MSS. The TM focuses the incident radiation directly onto the detectors themselves. One further feature of the TM is the degree of precision of the data. The quantization range of the MSS is sixty-four levels so that any one pixel can take a value in the range 0–63 inclusive, whereas the TM has 256 levels so that a pixel can take a value in the range 0–255 inclusive.

Landsat 4 suffered a number of problems in 1982 and 1983 which limited

*TABLE 5.7: Characteristics of the Multispectral Scanner (MSS) and Thematic Mapper (TM) carried on Landsats 4 and 5*

| *Band* | *MSS wavelength range (μm)* | *TM wavelength range (μm)* | *Principal applications for TM data* |
|---|---|---|---|
| 1 | 0.5–0.6 | 0.45–0.52 | Coastal water mapping<br>Soil/vegetation differentiation |
| 2 | 0.6–0.7 | 0.52–0.60 | Green reflectance by healthy vegetation |
| 3 | 0.7–0.8 | 0.63–0.69 | Chlorophyll absorption for plant species differentiation |
| 4 | 0.8–1.1 (1.0) | 0.76–0.90 | Biomass surveys |
| 5 | | 1.55–1.75 | Vegetation moisture<br>Snow/cloud discrimination |
| 6 | | 10.4–11.7* | Thermal mapping including plant stress |
| 7 | | 2.08–2.35 | Vegetation moisture and geological mapping |
| Ground pixel size | 82m | 30m (Bands 1–5, 7)<br>120m (Band 6) | |
| Quantization levels | 64 | 256 | |
| Data rate | 15 Mbps | 85 Mbps | |
| Weight | 68kg | 258kg | |
| Size | 0.35 × 0.4 × 0.9m | 1.1 × 0.7 × 2m | |
| Power | 50 watts | 332 watts | |

*Sources*: Freden and Gordon (1983), National Remote Sensing Centre (1985). See also Markham and Barker (1985a, b)

*Prelaunch calibration showed this wavelength range. The design range was 10.4–12.5μm

its performance. There was a failure of one of the X-band data transmitters shortly after the launch, followed later by problems with the solar panel cables. This led to the early launch of Landsat 5, particularly to provide TM coverage. By 1986 Landsat 4 was operating just an MSS and Landsat 5 both a TM and an MSS.

### *Data transmission*

The problems of data collection outside the range of US ground receiving stations have been partly overcome with Landsats 4 and 5, not by using the rather unreliable tape recorders carried on board Landsats 1–3, but by using the new Tracking and Data Relay Satellite System (TDRSS). The initial plan was for two TDRSS satellites to be placed in geostationary orbit above the equator on opposite sides of the globe. The new data route for Landsat was to transmit its data out to one of these TDRSS satellites which would then transmit the data to the White Sands receiving station in New Mexico. The data are then routed via the US Domsat domestic communications satellite to NASA's Goddard Space Flight Center in Maryland for processing. At present only one TDRSS satellite is in orbit, TDRSS-East at 41° W longitude which permits coverage of North and South America, Europe and Africa. The second TDRSS satellite was destroyed in the Space Shuttle Challenger explosion in January 1986. A second satellite will be launched by the Shuttle at some future date and will be positioned over 171° W longitude. This dual satellite system will then permit coverage of all of the globe, except for a small sliver of an area from the southern USSR, through Pakistan and India to the southern Indian Ocean which is a communications void for TDRSS.

Some of the receiving stations shown in Figure 5.7 have been upgraded to receive TM data, and large volumes of these data have been received at these stations for their own regions.

### *Commercialisation*

The development of Landsat and its sensors has been the work of NASA. NASA has a remit to undertake research and development but not to administer operational satellite systems. As a consequence NOAA took over operational responsibility for MSS operations in September 1982 and for TM operations in September 1984. Also in 1984 the US Department of Commerce issued a request for proposals for the commercialisation of Landsat. In June of that year the US Congress passed the *Land Remote Sensing Commercialisation Act of 1984* to enable the transfer of Landsat from the public to the private sector. Seven organisations submitted proposals to take over Landsat, and of these it was Eosat, a joint venture of Hughes and RCA, which was successful in its bid. With the signing of the contract on 27 September 1985 NOAA assigned operational control of Landsats 4 and 5 to Eosat, the Earth Observation Satellite Company, with a government budget support of $295 million. Eosat will continue to operate Landsats 4 and 5, and will launch and operate Landsats 6 and 7.

Eosat is expected to promote the development of a mature market-place for Landsat data. This has not so far happened naturally, not least because the satellites have until recently been mainly experimental rather than operational. The volume of sales of Landsat data has been large in terms of the number of items (imagery, digital data, etc.), but not so in terms of revenue. Table 5.8 gives a summary of Landsat sales for the period 1979–84. Even the total sales revenue of $6.6 million pa is not enough to sustain a satellite operating company, and as a result Eosat has increased the prices of Landsat products. It remains to be seen whether a mature, value-added industry can be developed from the excellent base of Landsat data.

*TABLE 5.8: Landsat sales volume and revenue, 1979–84*

| *Fiscal Year* | *USA* | | *Rest of the world* | | *Total* | |
|---|---|---|---|---|---|---|
| | *Items* | *$* | *Items* | *$* | *Items* | *$* |
| 1979 | 172,756* | 2,642,473* | 43,137 | 795,351 | 215, 893 | 3,437,824 |
| 1980 | 159,205* | 2,893,639* | 42,202 | 1,307,376 | 201,407 | 4,201,015 |
| 1981 | 144,389* | 3,013,326* | 58,163 | 2,177,540 | 202,552 | 5,190,866 |
| 1982 | 120,458 | 3,422,753 | 49,703 | 2,502,223 | 170,161 | 5,924,976 |
| 1983 | 56,007 | 3,850,557 | 32,938 | 2,196,870 | 88,945 | 6,047,427 |
| 1984 | 38,337 | 3,964,186 | 33,778 | 2,672,922 | 72,115 | 6,637,108 |
| *Totals* | 691,152 | 19,786,934 | 259,921 | 11,652,282 | 951,073 | 31,439,216 |

*Source*: Modified from NOAA (1986)
*Includes estimated figures from Goddard Space Flight Center sales

## SPOT

France has had a long-standing involvement in space activities. On 22 February 1986 at 1.44 am UT she launched the first of four satellites termed Satellite Probatoire d'Observation de la Terre (SPOT) (SPOT Image 1986). This first SPOT satellite was launched from the Kourou Space Centre in French Guiana on board an Ariane rocket. SPOT 1 is a polar orbiting platform at an altitude of 825km and is equipped with a High Resolution Visible (HRV) multi-linear array sensor (see chapter 3). The HRV has two modes:

(1) Multispectral mode (XS) with a pixel size of 20m and three wavebands in the visible and near infrared:
    Band 1  0.50–0.59μm
    Band 2  0.61–0.68μm
    Band 3  0.79–0.89μm
(2) Panchromatic mode (P) with a pixel size of 10m and a broad waveband of 0.51–0.73μm.

The working programme for SPOT was begun in the evening of the launch day, and on 23 February 1986 the first images of Nice and central

Algeria were transmitted to the receiving station at Aussuguel-Issus near Toulouse. In the first six weeks of its test and acceptance operation SPOT transmitted 12,000 scenes to Toulouse, including images of 80 per cent of the world's capital cities. One of these is shown in Figure 5.9, part of a SPOT 1 panchromatic image of Montreal acquired on 27 February 1986. The contrast on this scene is particularly good because of a light snow covering in parts of the city. It appears that imagery of this type will be suitable for mapping at scales of 1:50,000 to 1:25,000 (Jaques and Lopez 1986).

One significant difference between SPOT and Landsat is that SPOT is able to view to the side of its orbital path. This off-nadir viewing at up to 27° from vertical has two important implications. Firstly, stereoscopic coverage can be obtained by imaging the same area on different orbits at different viewing angles. Secondly, although SPOT will pass over the same point on the surface only every twenty-six days, the off-nadir viewing means that imagery can be acquired of the same area on successive days and that the average imaging interval can be as little as 2.5 days for priority areas.

The centre of SPOT activities is the receiving station at Toulouse. Ground receiving stations have also been established in Sweden, Canada and Bangladesh, and arrangements have been signed with receiving stations in India, China and Saudi Arabia.

SPOT 1 is the first operational satellite for Earth surface monitoring and it is operated by a commercial company, SPOT Image. SPOT Image will launch three more SPOT satellites: SPOT 2 will be identical to SPOT 1, but SPOT 3 and SPOT 4 (scheduled for launch in the 1990s) will introduce a new visible and near infrared sensor with a low spatial resolution (c. 1km) for biomass monitoring (SPOT Image 1985). In addition, the HRV instrument will carry a 1.5–1.7μm channel for vegetation studies. Thus in one satellite will be a competitor for both Landsat TM and Noaa AVHRR, combining high and low spatial resolution sensors on one platform, and doing so in an operational programme which will provide users with a reliable supply of data for more than a decade.

Finally on SPOT, the satellite also forms the basis of the Samro military reconnaissance satellite which is a joint venture between France and West Germany.

## Seasat

So far in this chapter the emphasis has been on visible and infrared sensors. In 1978 a remote sensing satellite with a synthetic aperture radar (SAR) and

5.9 SPOT 1 panchromatic image of Montreal, 27 February 1986. The pixel size is 10m. (Copyright CNES 1986)

a high spatial resolution was launched. The Seasat satellite carried an L-band SAR operating at 23.5cm wavelength and although, as its name implies, it was primarily intended for oceanographic applications, it collected valuable data over the land areas as well.

Seasat was launched into a near polar 108° orbit at an altitude of 790km in June 1978. The synthetic aperture radar provided digital image data with a spatial resolution of 25m and a swath width of 100km. This high spatial resolution was combined with the ability of the SAR to image the surface through clouds (see chapter 3) and so its data collection capabilities were not affected by weather. Figure 5.10 shows an example of a Seasat SAR image of southern Los Angeles obtained on 12 August 1978 at 04.09 GMT. The road network, the airport on the left (west) side and the harbour facilities near the bottom of the image are all clearly discernible and illustrate the potential for great spatial detail which SAR offers.

Seasat carried four other sensors in addition to the SAR (National Remote Sensing Centre 1985). These sensors are briefly described below.

(1) Radar altimeter to determine the sea surface height profile and wave heights.
(2) Radar scatterometer to measure wind speeds and directions.
(3) Microwave radiometer to measure sea surface temperature, rain rate and water vapour content of the atmosphere.
(4) Visible and infrared radiometer to image ocean and coastal features and measure sea surface temperatures.

The radar altimeter provided very interesting data on the topography of the ocean surface, with a relative height accuracy of up to 10cm. It then became clear that the contours of the ocean surface topography were a function of ocean bottom topography and so by using Seasat data a map of the ocean floor could be created.

Seasat's operation lasted only 106 days as the electrical system failed on 10 October 1978. However, despite its short life Seasat demonstrated the utility of spaceborne radar and prepared the way for radar imaging sensors in the 1980s and 1990s.

## NASA Space Shuttle

NASA's Space Shuttle was designed as a re-usable platform for experiments and data collection from space. As such it is an appropriate platform for satellite remote sensing, primarily for testing new sensor systems before an

5.10 Seasat L-band SAR image of Los Angeles, 04.09 GMT, 12 August 1978. The pixel size is 25m. (Courtesy MacDonald Dettwiler Associates)

operational system is constructed and launched. The Shuttle is normally launched into a near Earth orbit of c. 50° inclination because of the limitations of the launch site at the Kennedy Space Center in Florida. If the Shuttle orbit were to be near polar then the spacecraft would have to set off northwards over the populous eastern seaboard of the USA or southwards over the West Indies and South America. As the Shuttle launch design involves solid fuel booster rockets falling back to the surface after use then a northbound or southbound track over populated areas is clearly not desirable. When the Shuttle launch site is completed in California then the spacecraft will be able to be launched into a near polar orbit by launching it southwards over a long sea track above the Pacific Ocean.

January 28 1986 is fixed into the minds of all those associated with the NASA Space Shuttle. On that day the Space Shuttle Challenger exploded shortly after launch killing all seven crew. The cause of the explosion appears to be associated with fuel escaping from the solid fuel rocket boosters on either side of the Shuttle. The whole Shuttle programme was immediately grounded and Shuttle launches of all satellites deferred.

## SHUTTLE IMAGING RADAR

As the Seasat SAR had provided such interesting image data in 1978 there was an enthusiasm amongst US scientists, particularly those at the Jet Propulsion Laboratory in California, to take the opportunity of another space flight of a SAR as soon as possible. This was achieved in 1981 with the inclusion on the Space Shuttle of the first version of the Shuttle Imaging Radar, SIR-A. Like Seasat, indeed it was built using some of the same components as Seasat, SIR-A operated at L-band (23cm wavelength), it had a spatial resolution of 40m and a swath width of 50km. SIR-A collected data using an on-board optical recorder and no digital data were collected from this mission. In the eight hours which SIR-A operated it collected data for 10 million km$^2$ of the Earth's surface (Ford *et al.* 1982). An example of a SIR-A image is shown as Figure 3.5 in chapter 3. The imagery has been used for a variety of applications in geology, geomorphology, soil and agricultural studies, but one important application is the ability of radar to penetrate arid surfaces up to 2m depth. Sub-surface drainage networks beneath the Sahara desert were revealed for the first time on SIR-A images (ESA 1985).

The second imaging radar in the planned sequence of four instruments was SIR-B, flown on the Space Shuttle in October 1984 (Elachi *et al.* 1986). Again this SAR used L-band, it had a spatial resolution of 25m, and importantly it had both digital and optical recording. Unfortunately the data quality of SIR-B has not been uniformly good and the digital data of the UK, for example, are disappointing. NASA plans to fly improved versions of the radar on the Shuttle as SIR-C in 1990 and SIR-D some time later. SIR-C will have three frequencies (X, C and L band) and both like- and cross-polarisation (Elachi *et al.* 1986).

## METRIC CAMERA

In November/December 1983 the Space Shuttle carried Spacelab 1, the European Space Agency's contribution to NASA's Space Shuttle programme. Along with other experiments, Spacelab carried a modified Zeiss RMK A30/23 aerial survey camera which comprised the Metric Camera Experiment. The objective of the experiment was to explore the use of spaceborne cameras for topographic mapping at a scale of 1:50,000. The camera produced colour infrared black and white photography in 23cm × 23cm format, which from an altitude of 250km gave an image scale of 1:826,000 and a spatial resolution of c. 20m. Successive images taken in the flight direction had an overlap of 60 per cent to enable stereoscopic imaging. The thousand photographs taken by the Metric Camera showed that it was possible to prepare maps at 1:80,000 scale with a positional accuracy of 8m and an elevation accuracy of 13–27m. The best images were obtained in the arid sub-tropics, but over Europe results were not so good as the light levels in November and December were too low for optimum imaging.

## LARGE FORMAT CAMERA

The US competitor to the Metric Camera in spaceborne aerial photographic survey is the Large Format Camera (LFC) which was flown on the Space Shuttle in October 1984 at the same time as SIR-B. The LFC film format is 46 × 23cm and the camera is capable of a maximum spatial resolution of 5m. More details of the LFC are given in chapter 3.

## MOMS

The Modular Opto-electronic Multispectral Scanner (MOMS) was flown on the Space Shuttle in June 1983 and February 1984 to test the capabilities of an optical multi-linear array sensor. Each line of 6912 pixels was exposed simultaneously by four CCD elements. MOMS had two channels: visible (0.575–0.625μm) and near infrared (0.825–0.975μm) and a spatial resolution of 20m. A limited amount of high resolution imagery was gathered from the tropical parts of the globe.

## Costs

It is often difficult to tie down the costs of a remote sensing satellite because of the role of development costs and the extent to which they are written off by the manufacturer. However, in Table 5.9 there is an attempt to list the costs associated with six remote sensing satellite programmes.

*TABLE 5.9: Costs associated with six remote sensing satellite programmes*

| *Satellite programme* | *Cost (year)* | *Design life* | *Manufacturer* | *Remarks* |
|---|---|---|---|---|
| GOES 7 and 8 | $100 million (1984) | 5 years | Hughes Aircraft | Two spacecraft |
| Meteosat | $130 million (1984) | 3 years | Aérospatiale | Three spacecraft plus one spare |
| Insat | $130 million (1984) | 7 years | Ford Aerospace | Two spacecraft, two launches, ground station |
| Tiros-N/Noaa | $43.5 million (1984) | | RCA Astro-Electronics | Noaa 9 |
| DMSP | $171 million (1983) | 3 years | RCA Astro-Electronics | Four Block 5D-2 satellites |
| Landsat | $77 million (1978) | 3 years | General Electric | Landsat 4 |

*Source*: *Flight International* 12 January 1985

*Chapter 6*

# Digital image processing

## Why digital image processing?

Satellite remote sensing data are on the whole not acquired as pictures. Aside from the short-lived specific missions of SIR-A, the Metric Camera and the Large Format Camera, the majority of satellite image data are transmitted to Earth as digital values of the response of each pixel in the wavelength employed in the sensor. Only at a later stage are the digital values arranged in a two-dimensional image form and plotted as a picture. Thus the data have to be handled in digital form before an image can be produced, and once an image is produced then if it is in digital form further manipulations of the data can readily be performed by computer to extract useful information from the data.

Satellite remote sensing produces very large quantities of digital data. A single Landsat TM scene covering a ground area of 170km × 185km contains 273 Mbytes of data and occupies seven magnetic tapes when written at a tape density of 1600 bits per inch. Each SPOT HRV multispectral scene covering a ground area of 60km × 60km contains 27 Mbytes of data, and the SPOT operators have confirmed that the data supply is assured for over a decade. Noaa satellites can produce over 2,500 Mbytes of data from the AVHRR each day, although fortunately they do not do so. Clearly there is a data mountain in satellite remote sensing, and equally digital processing is the only sensible way of handling these vast quantities of information about our environment.

Human beings are good at interpreting images, but our eyes restrict us to looking in black and white at one band of data or in colour at a combination of three bands of data. With Landsat TM data there are seven bands available and with Noaa AVHRR five bands. It is only possible for human beings to look at a selection of any three of these at one time in a colour combination, so the remaining bands are not used. Human eyes cannot divide one picture by another, but the ratio of the near infrared to red reflectance provides useful information on vegetation. Many of the techniques discussed in this chapter are only possible by using digital image analysis and could not be achieved by visual interpretation alone. This is not to suggest that visual interpretation of satellite imagery is not useful: it is, and using conventional procedures of interpreting image tone, colour, size, shape, texture and context useful analyses of satellite images can be obtained.

## Context

### EQUIPMENT AND DATA STRUCTURE

In recent years a number of companies have developed image processing systems attached to standard computers. These so-called turnkey systems provide the hardware and the software ready for image processing as soon as the machine is installed and switched on, and typically consist of four main components:

(1) Input of digital data using a magnetic tape drive as most digital remote sensing data are published on computer-compatible tapes (CCTs).
(2) Storage of data on disk for fast access.
(3) Processing of the pixel data using the software provided.
(4) Display of the data and of the analyses performed on the data on a colour monitor.

In addition, a permanent form of the display results may be recorded using a camera attached to the colour monitor or by a laser plotter which creates images on transparent film. The image processing system is attached to a host computer such as the DEC PDP 11 or VAX series. A list of some of the turnkey systems suitable for satellite remote sensing applications is given in Table 6.1.

The list of four components above assumes that the data are available in digital form on CCTs. If only image data are available, say for SIR-A which was only equipped with an optical recorder, then digital data can be acquired using a scanning microdensitometer or video digitiser. The scanning microdensitometer has a rotating drum onto which a transparency of the image is fixed. Successive lines of the image are scanned by the drum

*TABLE 6.1: Examples of turnkey image processing systems employed in satellite remote sensing*

| *Name* | *Manufacturer* | *Address* |
|---|---|---|
| DIAD | DIAD Systems | Edenbridge, Kent, UK |
| Dipix | Dipix | Ottawa, Canada |
| GEMS | GEMS of Cambridge | Carlyle Rd, Cambridge, UK |
| IDIMS | Electromagnetic Systems Laboratories | Sunnyvale, California, USA |
| $I^2S$ | International Imaging Systems | Milpitas, California, USA |

rotating the transparency in front of a spot microdensitometer which measures the film's density at each point along a scan line. The process is repeated for all the scan lines of the image, and thereby produces a two-dimensional array of pixels in a form similar to that produced by a scanning radiometer. The pixel data are written in sequence onto a CCT for later analysis. A video digitiser is a television camera which digitises the density of each picture point as it scans an image placed in front of it. The digitised data are stored and then written sequentially onto a CCT.

From now on in this chapter it will be assumed that satellite remote sensing image data are available in digital form on computer-compatible tapes. The structure of the data is assumed to be a two-dimensional array of N rows of pixels and M pixels along each row. There may be only one array of data if only one band is used, or a series of similar arrays of data if more than one band is used. An illustration of this data structure is shown in Figure 6.1

## PATTERN RECOGNITION

The digital image processing techniques discussed here and implemented in image processing systems such as those listed in Table 6.1 are not designed solely for use with satellite remote sensing data, but form part of the wider field of pattern recognition (Gelsema and Kanal 1980, Kanal and Rosenfeld 1981). Pattern recognition has been neatly defined by Tou and Gonzalez (1974) as 'the categorisation of input data into identifiable classes via the extraction of significant features or attributes of the data from a background of irrelevant detail'.

This approach is relevant to data of many forms other than two-dimensional images, and pattern recognition techniques have been used in

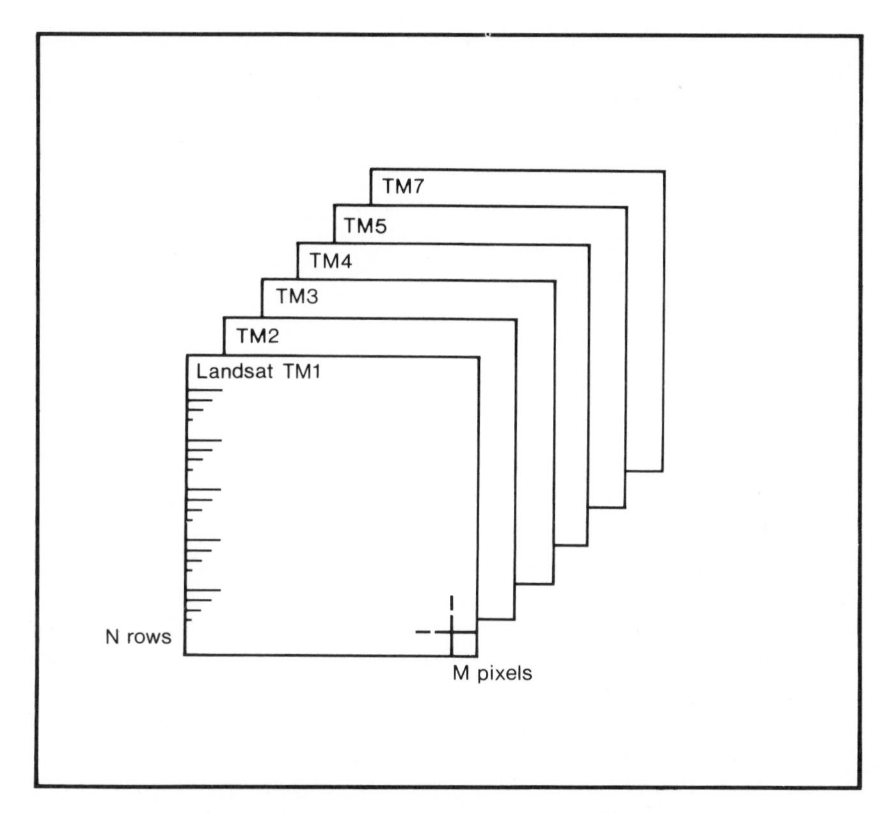


6.1 An illustration of the structure of Landsat Thematic Mapper digital data in a series of two-dimensional arrays, one array for each TM band

speech analysis, medical diagnosis, character recognition and materials testing as well as in satellite remote sensing (Kittler *et al.* 1981).

In satellite remote sensing the essence of the pattern recognition task is the identification of the desired features, classes or categories, such as wheat fields, eroded soils or cloud types, from the mass of information which comprises the digital image. This identification can be achieved using the techniques described in this chapter.

## Initial analysis

### PREPROCESSING

Digital image data, when first received from a satellite, contain a number of distortions which must be corrected. Two forms of correction are required:

geometric correction to correct for changes in the shape of the image, and radiometric correction to correct for variations in the radiance received across a scene (Duggin *et al.* 1985). Increasingly the archives of satellite remote sensing data, such as those at the US EROS Data Center, the UK National Remote Sensing Centre and the French SPOT archive, hold data which have been both geometrically corrected, for example to a Mercator projection or to the British National Grid, and radiometrically corrected in some way (Anderson 1985, Bryant *et al.* 1985, Hardy 1985, Welch *et al.* 1985). Because of this the two groups of corrections are discussed here in general rather than in specific terms.

### *Geometric correction*

The principal types of geometric distortion are shown in schematic form in Figure 6.2. These distortions are caused by variations in the Earth's rotation, variations in the velocity or altitude of the satellite, and variations

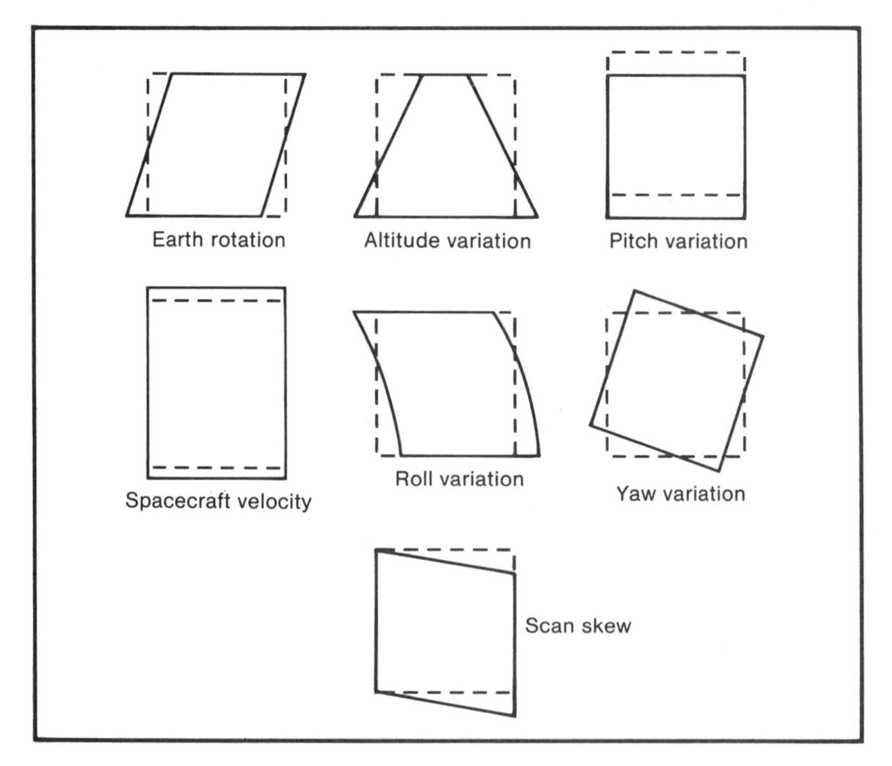


6.2 Geometric distortions in satellite remote sensing data. The dashed lines indicate the desired image shape, the solid lines the actual image shape. (After Bernstein and Ferneyhough 1975, Sabins 1978)

in the pitch, roll or yaw of the satellite platform. Geometric correction to account for these variations has been achieved in one of two principal ways. Firstly, by modelling the spacecraft orbit geometry it is possible to predict what the satellite should be viewing and so the real image can be modified accordingly. This is the approach adopted by the Environmental Research Institute of Michigan in the USA. Secondly, by the use of ground control points (GCPs), where GCPs are identified on both the image and on a map at the required projection: the GCPs may be road intersections, airport runways, stream junctions, coastal spits, jetties, etc. (Benny 1983, 1985, Davison 1986, Orti *et al.* 1979). Once a sufficient number of GCPs (50–150) has been identified scattered about the image and the map then an equation can be calculated which relates the row/column pixel coordinates of the image GCPs to the coordinates of the map GCPs. This equation can then be applied to all the pixels of the image to re-map them onto the map coordinate system. This process is analogous to warping the image to fit the map. A discussion of the cartographic requirements of satellite remote sensing data is given in chapter 7.

### *Radiometric correction*

Radiometric corrections are not as simple as geometric corrections and are more sensor and scene dependent. Multiple detectors in the sensors such as the Landsat MSS or TM have slight physical variations which mean that there will be variations in the radiance recorded by different detectors even if they are looking at the same object. In Landsat MSS data there is an effect termed six-line banding whereby in an image there is often clearly a striping effect repeated down the image every six lines. This is because the Landsat MSS scans six lines simultaneously in its four wavebands and the energy from each scan line is directed to a different detector (see chapter 5 and Figure 3.2). Each detector behaves slightly differently so a repetitive six-line banding effect is produced. In Landsat TM data there is a sixteen-line banding effect for a similar reason: the TM scans sixteen lines simultaneously and directs the energy onto different detectors. The TM also scans in both directions across the ground track of the satellite, and this too can lead to banding (Fusco *et al.* 1986). These banding or striping effects can be removed by filtering the image data.

The SPOT HRV has only one line of detectors so a comparable problem to that with Landsat MSS and TM does not occur. However, the HRV instrument consists of separate CCD elements in its multi-linear array which all have to be calibrated to a common standard before the satellite is launched.

Atmospheric effects such as haze, water vapour and sub-pixel sized cloud all reduce the radiance arriving at a sensor, particularly at the shorter wavelengths, and so the effects of the atmosphere have to be estimated (Duggin and Saunders 1984, Duggin and Philipson 1985). In sensors with wide scan angles, such as Noaa AVHRR and Meteosat, the distance which energy travels to the sensor from objects on the edge of the scan line is greater than the distance from objects in the centre of the scan line (nadir). In particularly, the distance through the atmosphere is greater and results in limb darkening, that is lower radiance levels recorded at the edge or limb of the scan than at the centre. This is corrected either by omitting the outer parts of the scan lines or by modelling the decline in radiance towards the edges of the scan.

## DENSITY SLICING

Each pixel in a one-band image commonly takes a value in the range 0–255, depending on the sensor system involved. If groups of densities in this range can be associated with certain environmental features then a map can be created by separating out the groups of densities and assigning colour codes to them. This is termed density slicing, and can best be illustrated by taking a single scan line across a Landsat image.

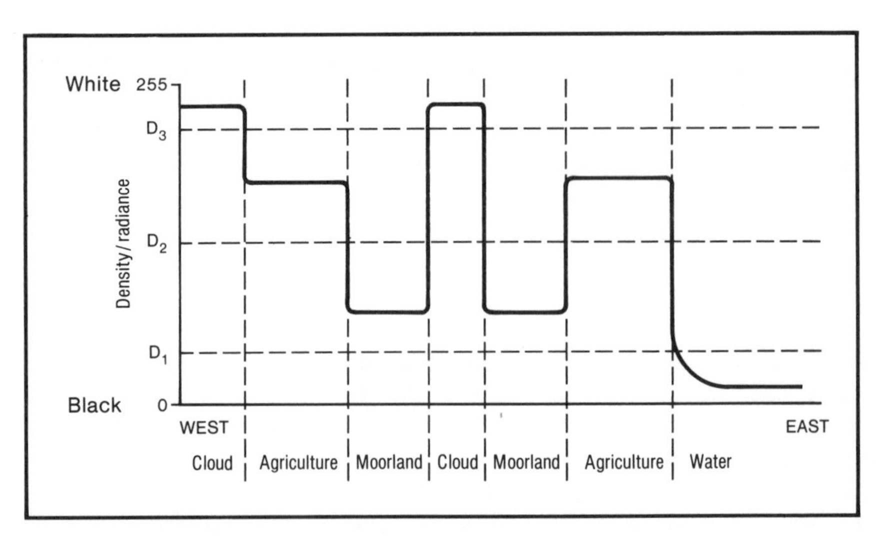


6.3 Simplified density or radiance variation of a Landsat TM4 scan line across the North York Moors

Figure 6.3 shows in simplified form the density or radiance variation across a single line of a Landsat Thematic Mapper band 4 near infrared image of part of the North York Moors. Each terrain cover type is associated with a distinct density range across this scan line, and from brightest to darkest the order is: cloud, agriculture, moorland, water. If three density values are identified ($D_1$, $D_2$, $D_3$) then this provides a means of slicing the whole density range 0–255 so that groups of densities are associated with certain terrain cover types. The groups are given in Table 6.2. This procedure is repeated for all the scan lines in the image, so producing a map of densities grouped together and related to terrain cover types. The density slices can then be colour coded using an image display, perhaps using the colour code scheme included in the right-hand column of Table 6.2.

*TABLE 6.2: Terrain cover types associated with the density slices in Figure 6.3*

| *Density slice* | *Terrain cover* | *Colour code* |
|---|---|---|
| Densities below $D_1$ | Water | Dark blue |
| Densities between $D_1$ and $D_2$ | Moorland | Brown |
| Densities between $D_2$ and $D_3$ | Agriculture | Green |
| Densities above $D_3$ | Cloud | White |
| Densities between $D_1$ and $D_3$ | Land | |

This simple procedure can be applied to a wide variety of digital imagery when the densities can be associated with certain terrain cover properties. This is a useful technique, although there are limitations to it, and using just one band of digital imagery may use only one part of the information base.

## BAND RATIOS

Two bands of data can be combined together by addition, subtraction or, more commonly, division of the pixel radiances. As was seen in chapter 2 green vegetation has a high near infrared reflectance and a low red reflectance, and by contrast bare surfaces do not show such sharp distinctions (see Figure 2.6). A common band ratio is to divide the near infrared reflectance by the red reflectance (Tucker 1979), which gives high values for healthy green vegetation and low values for bare areas. As illustration the relevant band combinations for four satellite sensors are shown in Table 6.3.

NOAA produces an operational product for the northern and southern hemispheres which is a combination of band subtraction and band ratio, the

Normalised Vegetation Index (NVI) (Yates et al. 1986). The NVI is calculated as shown in equation 6.1 (Schneider *et al.* 1985).

$$\text{NVI} = \frac{\text{AVHRR2} - \text{AVHRR1}}{\text{AVHRR2} + \text{AVHRR1}} \tag{6.1}$$

and the data are described in chapter 5. The use of the NVI is described in chapter 7.

*TABLE 6.3: Example band ratio combinations for four satellite sensors. The numerator is the relevant near infrared band and the denominator the red*

| *Satellite sensor* | *Numerator* | *Denominator* |
|---|---|---|
| SPOT HRV | XS3 (0.79–0.89μm) | XS2 (0.61–0.68μm) |
| Landsat 5 TM | TM4 (0.76–0.90μm) | TM3 (0.63–0.69μm) |
| Landsat 5 MSS | MSS4 (0.8–1.1μm) | MSS2 (0.6–0.7μm) |
| Noaa AVHRR | AVHRR2 (0.725–1.1μm) | AVHRR1 (0.55–0.68μm) |

A number of other indices have been proposed in the literature including the transformed vegetation index, the environmental vegetation index and the vegetative–sponge index, and some of these are reviewed by Hayes (1985). Short (1982, page 167) gives illustrative examples of the application of four such indices to a Landsat image of an area around Blue Mountain, Pennsylvania.

Wardley and Curran (1984) discuss the use of the Perpendicular Vegetation Index (PVI). This index is perpendicular in the sense that it defines for each pixel the distance that it is from the plane of soils in the feature space defined by the near infrared and visible wavebands. In Figure 2.8 in chapter 2 there are two points plotted in the vegetation zone. The perpendicular distance of each point from the plane of soils line is its PVI value. The PVI is calculated as shown in equation 6.2 (Wardley and Curran 1984)

$$\text{PVI} = \sqrt{[(R_s - R_v)^2 + (NIR_s - NIR_v)^2]} \tag{6.2}$$

where R is the reflectance in the red waveband to represent visible reflectance, NIR is the reflectance in the near infrared waveband and subscripts s and v refer to the soil background and the vegetation respectively. Equation 6.2 therefore measures geometrically how far a plotted point is away from the plane of soils. If a pixel contains only bare soil its PVI value will be zero as $(R_s - R_v)$ and $(NIR_s - NIR_v)$ will both be zero: as the vegetation amount increases then PVI increases.

### PRINCIPAL COMPONENTS ANALYSIS

The discussion of band ratios essentially concentrates on two wavebands chosen in the visible and the near infrared. A larger number of bands can be combined together by using principal components analysis (PCA), and this technique has been used extensively with Landsat MSS and TM data. Principal components analysis involves finding mathematically a new set of variables which describe the variance in the original data set, but which are independent of each other, i.e. the correlation between the new variables is zero. These new variables, or principal components, are mathematical combinations of the original wavebands. The first principal component accounts for the largest proportion of the variance in the original data set, the second principal component the next largest proportion of the variance, and so on. The first principal component normally shows albedo or brightness for visible and near infrared data such as Landsat MSS, and the second principal component often shows an indication of greenness. However, the use of PCA is always data dependent and the components produced must be interpreted in the light of the raw data. Townshend (1984) and Townshend *et al.* (1983) give examples of principal component images generated from Landsat TM and airborne scanner data.

As the principal components are independent of one another then a colour combination of the first three components can be useful in providing maximum visual separability of image features. These first three components can be displayed as each of red, green and blue on a colour monitor, and this often produces a better colour image than the use of the original wavebands (Canas and Barnett 1985).

## Multispectral analysis and classification

Different surfaces have different spectral reflectance curves (see Figure 2.6). If the *spectral* reflectance at *many* points on such curves (viz. multispectral) can be extracted then the different surfaces can be characterised and quantitatively compared by their spectral characteristics. In practice this involves plotting radiance values in the feature space defined by the wavebands employed. In the case of two wavebands the feature space is two-dimensional, as shown in Figure 6.4 for red and near infrared reflectance. As the number of wavebands increases so the dimensions of the feature space increase accordingly. For example, a multispectral analysis of Landsat TM data may use the six non-thermal bands and so the feature space will be six-dimensional. Computationally this presents few problems but in diagram-

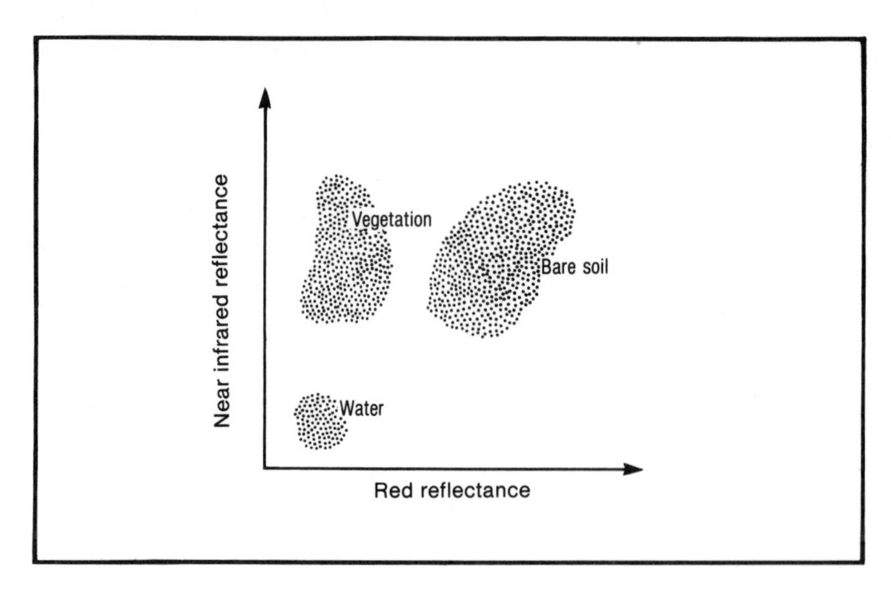


6.4 Feature space defined by near infrared and red reflectance. Example features of vegetation, bare soil and water are plotted in the space defined by the two axes

matic terms we are constrained to drawing in three dimensions. For most practical purposes two-dimensional diagrams are clearer, so in this section the two-dimensional case will be shown, although the principles are readily extendable to greater dimensions.

## SAMPLE DATA

For most multispectral analyses a set of sample data needs to be extracted from the image data. These sample data are often termed training data as the classifier is trained on these data and uses them to calculate the classification model parameters. The choice of the training sample data should consider the following points:

(1) The location of the sample pixels should be chosen using some form of random sampling scheme.
(2) The number of training samples should be at least ten times the number of wavebands used.
(3) The sample data should ideally have a multivariate normal distribution.

Once the training data have been identified then they may be used with one of the classifiers discussed below. The purpose of each classifier is briefly described; for a mathematical treatment see Swain and Davis (1978).

## BOX CLASSIFIER

The simplest form of multispectral analysis is to define a box in the feature space which identifies the range of one cover type. This can be done in one of two ways: qualitatively and quantitatively. Figure 6.5 shows a qualitative example where cloud types can be identified from a combination of their brightness in visible and thermal infrared wavebands. The descriptors of image brightness (dark, dim, etc.) can be converted to numerical ranges for a particular application and to more than two wavebands. Liljas (1982) illustrates the use of visible, near infrared and thermal infrared wavebands to identify cloud types in a three-dimensional box classification scheme.

A quantitative box classifier is illustrated in Figure 6.6. Here the range of values plotted on each axis for any one cover type is defined as the mean

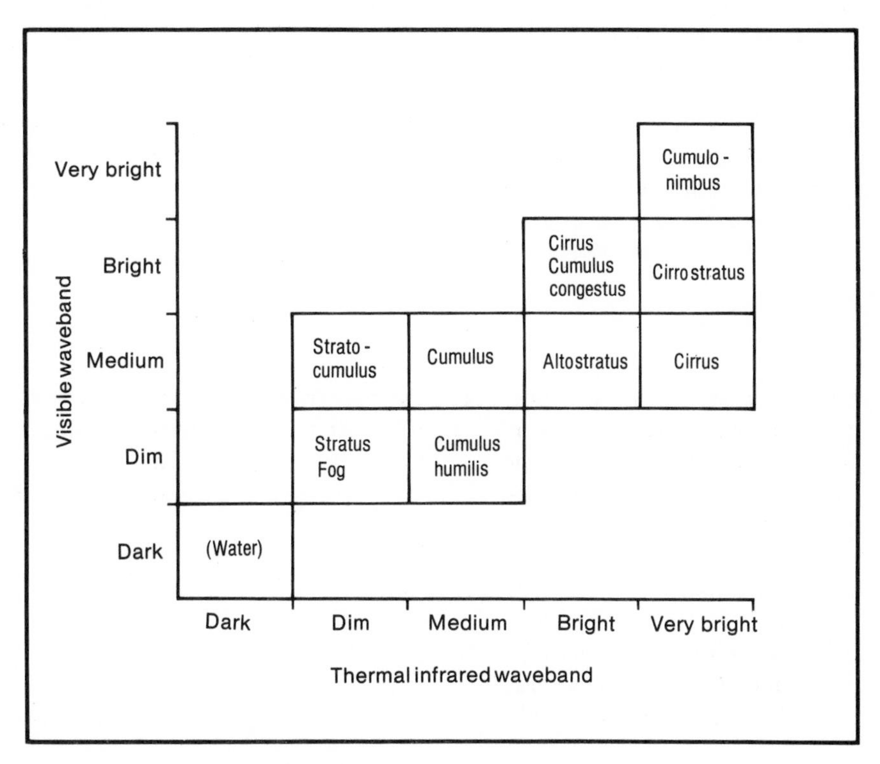


6.5 Qualitative box classifier for cloud type recognition

minus one standard deviation to the mean plus one standard deviation. These limits on both axes define a box for each terrain cover type in the feature space. Boundaries other than one standard deviation may be chosen, depending on the statistical behaviour of the data and the experience of the user with the data.

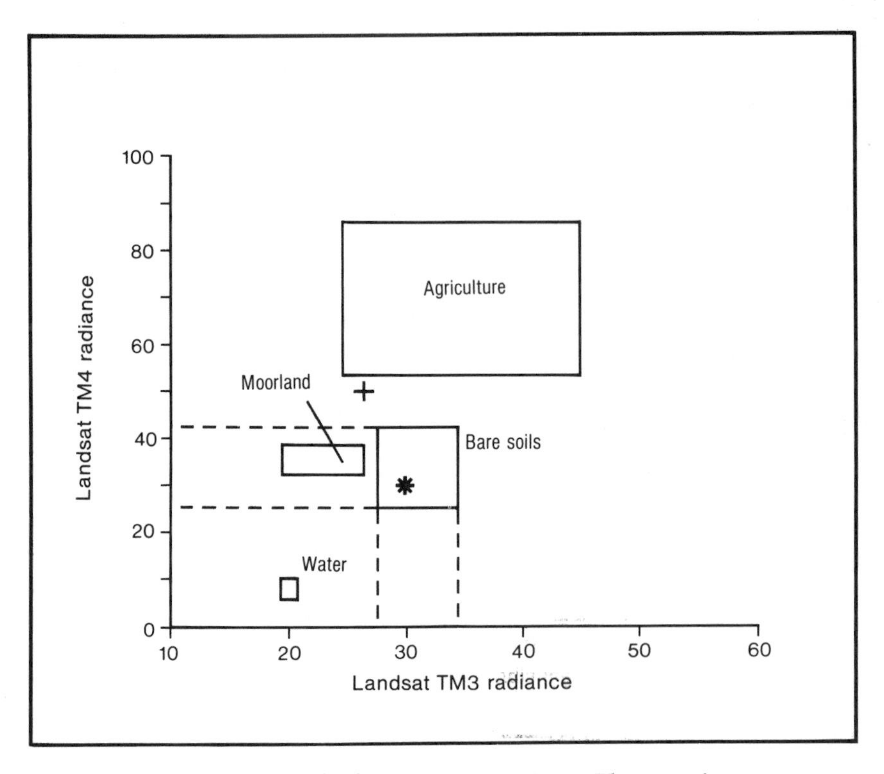


6.6 Numerical box classifier for four terrain cover types. The asterisk is an unknown pixel classified as bare soil. The cross is an unknown pixel not classified as it falls outside all the boxes

Having defined these classification boxes then the values for an unknown pixel can be plotted in the feature space. If a point falls inside a box then it is classified as belonging to the class of that box. An example of an unknown point, marked by an asterisk, is shown in Figure 6.6: it falls into the class bare soils and so is classified as belonging to that class. There will be unknown points plotted which do not fall into any box and so they remain unclassified; an example of this is marked by a cross in Figure 6.6.

## DISCRIMINANT ANALYSIS

The boxes shown in Figure 6.6 have regular boundaries at right angles to one another. This may not be the best way of delimiting the distribution of a class and so other boundary lines have to be sought. Discriminant analysis provides one means of separating two classes, and several discriminant analyses can be combined to divide up a feature space on an irregular basis. Figure 6.7(A) shows two classes: agriculture and bare soil in the feature space defined by Landsat TM bands 3 and 4. Figure 6.7(B) shows the line which best separates these two classes. This line is determined mathematically by finding the one-dimensional projection onto which the two distributions can be plotted which separates the two classes the most. The discriminant function line is then drawn at right angles to this projection at a point which bisects the two distributions. This process can be repeated for any pair of classes, thereby constructing a more complex division of the feature space.

As with the box classifier, unknown points can be plotted in the feature space and their class determined by which side of the discriminant function line they fall. This is illustrated in Figure 6.7 with points marked by an asterisk and a cross which are classified as moorland and bare soil respectively. In the case of discriminant analysis the feature space is normally completely divided so there is no area of unknown class.

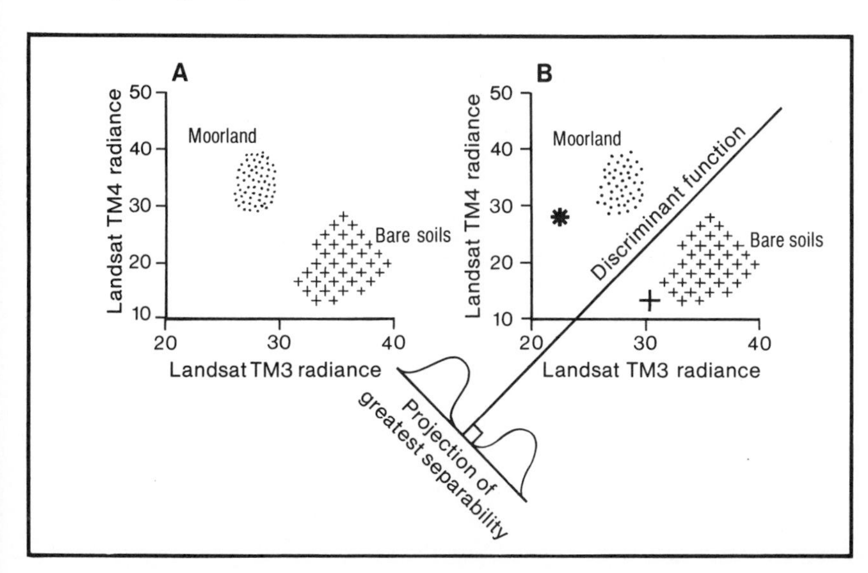


6.7 Discriminant analysis classifier. The asterisk is an unknown pixel classified as moorland. The cross is an unknown pixel classified as bare soil

## MAXIMUM LIKELIHOOD CLASSIFICATION

The maximum likelihood model determines the probability of all points in the feature space belonging to a particular class, so that each point has a probability of belonging to each of the classes identified. The class to which a point belongs is the class for which it has the highest probability, that is the maximum likelihood of belonging to that class. This can be illustrated diagrammatically in Figure 6.8 where around each class centre are probability contours which define the probability of belonging to that class. Unknown points can be allocated to a class by determining that class which has the highest probability, i.e. the most likely class to which it will belong. In Figure 6.8 an unknown point is again marked by an asterisk. It has the highest probability of belonging to the bare soils class, and so is allocated to that class.

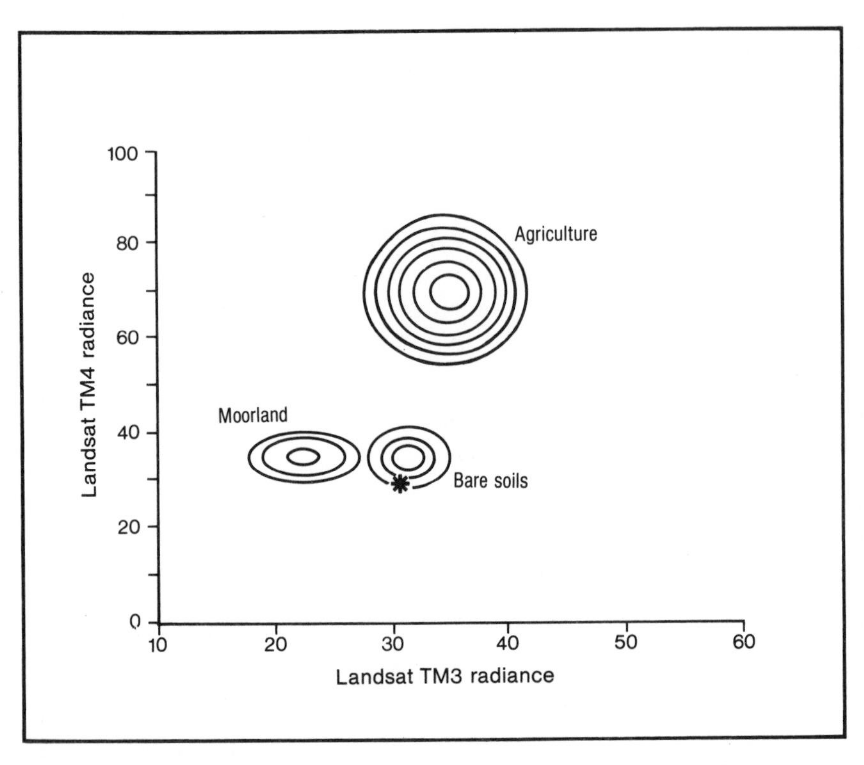


6.8 Maximum likelihood classifier showing equiprobability contours for three terrain cover types. The asterisk is an unknown pixel classified as bare soil

## CLUSTER ANALYSIS

The three techniques discussed so far employ data for known groups or classes in the training data. If the groups are unknown or not precisely known then what we require is a technique which will detect the inherent or natural structures in the data: cluster analysis attempts to do this. Clusters are identified in the feature space by defining cluster centres which (1) minimize the distances within a cluster of points, and (2) maximize the distance between the centres of the clusters. In Figure 6.9 the data for a set of unclassified sample pixels were plotted in the red/near infrared feature space of Landsat TM bands 3 and 4 and the cluster centres identified by a clustering algorithm (Swain and Davis 1978). On re-inspection of the data these cluster centres were then named (agriculture, bare soil, etc.) and so the classes can be identified. Classification is achieved for unknown points by plotting them in the feature space and assigning to them the class of the nearest cluster centre. In the case of Figure 6.9 the unknown point shown by

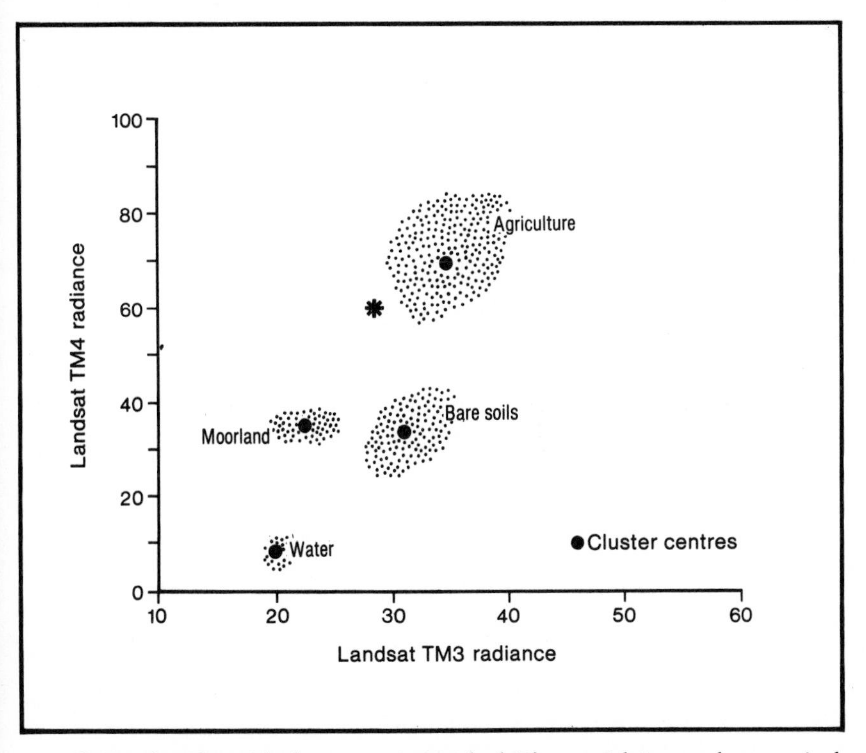


6.9 Cluster classifier with cluster centres marked. The asterisk is an unknown pixel classified as agriculture

an asterisk is nearest to the agriculture class cluster centre and so is classified as agriculture.

## Texture analysis

One feature of all the analyses above is that they are all applied to single point data. That is, they all extract pixels in isolation from the image data and calculate statistics from those pixel data. This gives no regard to the spatial relationships of the pixels. For example, the statistics calculated in discriminant analysis would be the same as long as the pixel data were the same no matter what spatial arrangement the pixels displayed: the pixels could be re-sorted randomly and the same statistics would be produced as when the pixels were extracted from a real image.

This lack of spatial information can be partly rectified by building into image analysis an assessment of image texture. Qualitatively we all think we know what texture is, but it is rather more difficult to put texture into precise terms and even more difficult to express it quantitatively. Most dictionary definitions of texture are extensions of descriptions related to weaving: smooth, rough, fine, coarse, twilled, ribbed, loose, plain, etc. These terms can be used to describe image texture qualitatively and so these are the terms which are useful when designing algorithms for analysing image texture quantitatively. However, the quantitative definition of texture is rather more difficult and most attempts have simplified, explicitly or implicitly, the definitions of texture to a range of textures from smooth (or no texture) to rough (or very textured).

Image texture varies from place to place on an image and it is this variation which enables us to use texture as an indicator of image features. For instance, water commonly has a smooth texture and a field of cumulus clouds or a woodland a rough texture on satellite imagery. The framework used to analyse this spatial variation is to divide up the image into small areas and to analyse the texture within each area. These are termed windows of the main image and their structure is shown diagrammatically in Figure 6.10. Such windows are $n \times m$ pixel sub-arrays of the main $N \times M$ pixel image array. The windows may be discrete or they may overlap, and are commonly between $3 \times 3$ and $12 \times 12$ pixels in size, although this will vary depending upon the application.

Having defined the structure of pixel windows covering the image the problem then becomes what to measure in each window to measure texture quantitatively. Two main approaches have been adopted in the literature: first order and second order texture measures (Weszka *et al.* 1976).

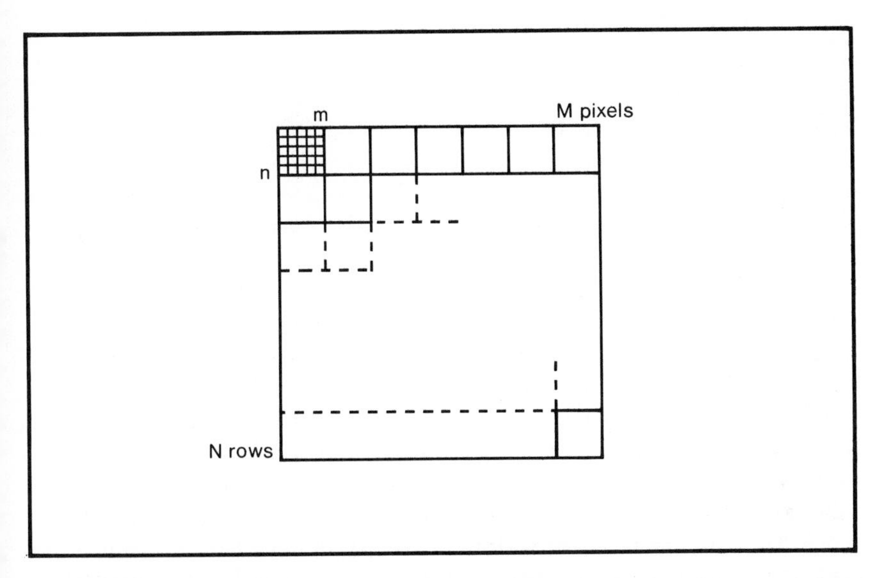


6.10 Schematic illustration of the partition of an image consisting of N rows and M pixels per row into windows or sub-arrays each of n × m pixels

## FIRST ORDER TEXTURE MEASURES

The first order measures of image texture operate on the pixels in each window directly. A very useful first order texture measure is the standard deviation (s):

$$s = \sqrt{\frac{\sum_{i=1}^{n} \sum_{j=1}^{m} (z_{ij} - \bar{z})^2}{(n \times m) - 1}} \tag{6.3}$$

where z is the radiance value of pixel ij, n is the number of rows in the window and m is the number of pixels in each row. If the value of each pixel in the window is the same then clearly the texture is smooth with no spatial variation and the value of s is zero because the difference between each $z_{ij}$ and the mean $\bar{z}$ is always zero. As the texture moves from smooth to rough then the value of s increases.

Other first order measures have included the coefficient of variation, measures of consistency, eigenvalues and spatial autocorrelation. Examples of these can be found in Thomas and Davey (1977), Harris and Barrett (1978) and Harris (1980).

## SECOND ORDER TEXTURE MEASURES

The second order measures of texture do not operate on the window pixels themselves, but on a matrix derived from the window pixels and so one stage removed from the raw data. This approach has been developed extensively by Robert Haralick (Haralick 1979, Haralick and Shanmugan 1974, Haralick *et al.* 1973).

The first stage of the process is to construct a new matrix which characterises the spatial dependence of the pixels in each window. This spatial dependence matrix is best shown by illustration in Figure 6.11. Figure 6.11(a) shows a simple image window of 8 × 8 pixels for which texture is to be calculated. This could be one of the image windows shown in Figure 6.10. The values of each pixel are in the range of 0–3 for simplicity. A new matrix (Figure 6.11(b)) is constructed whose rows and columns are the densities of the original image: in this example image densities 0, 1, 2 and 3. Each element of this new matrix is filled by the number of times that pixels with certain image densities occur next to one another in the original image (Figure 6.11(a)). In Figure 6.11(b) the matrix element 0–0 is filled by counting the number of times that density 0 falls next to density 0 in the original image window. In the horizontal case this is twelve times so the element 0–0 of Figure 6.11(b) is 12. Element 0–1 is the

**A**

| Rows \ Pixels | 1 | 2 | 3 | 4 | 5 | 6 | 7 | 8 |
|---|---|---|---|---|---|---|---|---|
| 1 | 0 | 0 | 1 | 1 | 0 | 0 | 1 | 0 |
| 2 | 0 | 0 | 1 | 1 | 1 | 1 | 0 | 1 |
| 3 | 0 | 1 | 1 | 1 | 1 | 1 | 0 | 0 |
| 4 | 1 | 1 | 2 | 2 | 2 | 1 | 1 | 0 |
| 5 | 2 | 2 | 2 | 2 | 2 | 2 | 2 | 2 |
| 6 | 2 | 2 | 3 | 3 | 3 | 3 | 3 | 2 |
| 7 | 3 | 3 | 0 | 0 | 0 | 0 | 3 | 3 |
| 8 | 3 | 0 | 0 | 0 | 0 | 0 | 0 | 3 |

**B**

| Image density | 0 | 1 | 2 | 3 |
|---|---|---|---|---|
| 0 | 12 | 10 | 0 | 4 |
| 1 | 10 | 10 | 2 | 0 |
| 2 | 0 | 2 | 10 | 2 |
| 3 | 4 | 0 | 2 | 6 |

6.11 (a) Image window of 8 × 8 pixels whose densities are in the range 0–3 (b) The spatial dependence matrix **P** calculated from the image in (a) for nearest horizontal neighbours

number of times density 0 falls next to density 1 in the image window: ten times. Element 0–2 is the number of times density 0 falls next to density 2 in the image window: none. All the elements of Figure 6.11(b) are completed using this simple counting method, and this creates a matrix **P** on which the second order texture measures are calculated.

In this simple example the original range of densities is only 0–3, which is rather unrealistic. In real applications the range of image densities is more likely to be 0–255, so the matrix **P** will be 256 × 256 elements in size. The pattern shown in Figure 6.11(b) comprises counting only for horizontal neighbours, but **P** matrices can also be calculated for vertical or diagonal neighbours or neighbours at more than one pixel removed.

Once **P** is produced then a number of statistics can be calculated from it; three of these are homogeneity, contrast and correlation. Each is designed to extract textural information from the spatial dependence information contained in matrix **P**. It is clear from the description that the method is more involved and more computationally complex than the first order texture measures which are calculated from the pixel values directly.

### *Texture and classification*

Once the texture statistics have been calculated for windows of the image then sample values can be incorporated into classifications. The texture analyses add extra information planes which extend the multispectral layers of information shown in Figure 6.1. The texture data can be included in the types of multispectral analysis described earlier in this chapter in much the same way that a new waveband might be incorporated. Instead of one of the axes of Figures 6.6–6.9 being the response at a particular waveband it is replaced by a texture axis.

# Assessment

## CLASS ACCURACY

Once a classification has been produced using the methods described above then it must be assessed for accuracy. Sample test points in each class are chosen, and the class assigned by the analysis of the satellite remote sensing data is compared with some form of reference data such as ground checking, maps, aerial photographs or other sensor data (Curran and Williamson 1985, Hubbard and Wright 1982). Although this sounds simple there are a number of issues involved, and some of these are described below.

(1) The choice of the sample test points should be done ideally according to a random sample design. This may not be possible in reality because of access limitations, but the sample points should not be located only at roadside locations where access is easy.
(2) The sample data used to design the classifier should not be used to assess the accuracy of the classifier.
(3) The number in the sample should be about fifty for each class, although this may be smaller if the class is only thinly represented on the image classification.
(4) The image classification and the reference data are arranged in a classification matrix, also termed an error matrix or confusion matrix (see Table 6.4). This matrix gives an indication of where the classification is performing well and where badly.

*TABLE 6.4: A sample classification matrix of five terrain cover classes*

| | | *Image analysis class* | | | | | |
|---|---|---|---|---|---|---|---|
| | | *Water* | *Forest* | *Pasture* | *Soil* | *Urban* | *Row total* |
| *Reference field data class* | Water | 48 | 2 | 0 | 0 | 0 | 50 |
| | Forest | 2 | 41 | 5 | 2 | 0 | 50 |
| | Pasture | 0 | 5 | 40 | 5 | 0 | 50 |
| | Soil | 0 | 0 | 2 | 42 | 6 | 50 |
| | Urban | 0 | 0 | 0 | 6 | 44 | 50 |
| | Column total | 50 | 48 | 47 | 55 | 50 | |

*Source*: After Curran (1985)

(5) It is tempting to calculate the overall classification accuracy as the number of pixels classified correctly divided by the total of those correctly classified plus those incorrectly classified. This would give a misleading result because some sample pixels will be correctly classified even in a random classification of pixels. For example, if forty-five of the fifty sample pixels of one class were classified correctly then it cannot be assumed that the classification accuracy is 90 per cent (Curran 1985). According to probability theory the accuracy of the classification of 45/50 at the 95 per cent level of significance will be in the range 83.8–95.2 per cent. This is illustrated in Table 6.5 for the classification matrix of Table 6.4. Hubbard and Wright (1982) give a similar example in the classification of ten surface types in Landsat MSS data of Scotland.

*TABLE 6.5: Classification accuracy for each class shown in Table 6.4*

| *Class* | *Proportion correct* | *Class accuracy range (%)* |
|---|---|---|
| Water | 48/50 | 86.5–98.9 |
| Forest | 41/50 | 69.2–90.2 |
| Pasture | 40/50 | 69.0–88.8 |
| Soil | 42/50 | 71.5–91.7 |
| Urban | 44/50 | 76.2–94.4 |
| Total | 215/250 | 81.2–89.8 |

*Source*: Curran (1985)

## DIVERGENCE

The classification of satellite remote sensing data is an exercise in reduction. It is also an exercise in separation because classes are sought which are clearly separable one from another. One way to assess the separation of classes is by means of a statistical measure of divergence which assesses the extent to which any pair of classes is separable one from another in a multispectral classification. In addition, statistical separability is related to the *a priori* estimate of the probability of correct classification (Swain and Davis 1978, Yool *et al.* 1986).

Singh (1984) gives a condensed summary of divergence as a measure of separability between classes. The pairwise divergence ($D_{ij}$) between a pair of classes i and j is given by

$$D_{ij} = \tfrac{1}{2} \text{ tr } [(\mathbf{C}_i - \mathbf{C}_j)(\mathbf{C}_j^{-1} - \mathbf{C}_i^{-1})] + \tfrac{1}{2} \text{ tr } [\mathbf{C}_i^{-1} + \mathbf{C}_j^{-1})(M_i - M_j)(M_i - M_j)^T] \quad (6.4)$$

where $\mathbf{C}$ is the class covariance matrix, $\mathbf{C}^{-1}$ is the inverse of the covariance matrix and M is the mean vector. T denotes the transpose of a matrix and tr the trace.

The range of $D_{ij}$ from equation 6.4 is from 0 to infinity, with larger values denoting greater separation. To provide a bounded range a transformed divergence ($TD_{ij}$) is defined as

$$TD_{ij} = 2000[1 - \exp(-D_{ij}/8)] \quad (6.5)$$

This expression has an upper limit of 2000, defined by the first constant in the expression, and the value for maximum separation between two classes i and j is 2000; a value for $TD_{ij}$ of zero implies no separability.

*TABLE 6.6: Transformed divergence values for ten pairs of terrain cover types extracted from a classification of Landsat TM data of Sidi Ali ben Aoun, central Tunisia. (Data courtesy Sinclair Sutherland)*

| | *Terrain class* | | | |
|---|---|---|---|---|
| *Terrain class* | *Bare rock* | *Scrub* | *Cereals* | *Other agriculture* |
| *(a) Input data are Landsat TM bands 1, 2, 3 and 4* | | | | |
| *Urban* | 1521 | 1436 | 1998 | 1808 |
| *Bare rock* | | 1749 | 2000 | 1981 |
| *Scrub* | | | 1990 | 1731 |
| *Cereals* | | | | 1042 |
| *(b) Input data are Landsat TM bands 1, 2, 3, 4, 5 and 7* | | | | |
| *Urban* | 1723 | 1944 | 1999 | 1869 |
| *Bare rock* | | 1955 | 2000 | 1992 |
| *Scrub* | | | 1994 | 1801 |
| *Cereals* | | | | 1090 |

Table 6.6 shows an example of transformed divergence values. The matrices show the values of $TD_{ij}$ between each of ten pairs of terrain classes extracted from a classification of Landsat TM data of central Tunisia. There is no critical value of $TD_{ij}$ which defines a boundary between those pairs which have clear separation and those which do not, but in practice a value of $TD_{ij}$ of 1700 is often taken as an indicator. In Table 6.6(a) there is clear separation of cereals from urban, bare rock and scrub because these three values are at or close to 2000. There is poor separation of urban from bare rock and scrub with values c. 1500, and as might be expected poor separation of cereals from other agriculture with a value of 1042.

The data used to compile Table 6.6(a) were Landsat TM bands 1, 2, 3 and 4, which are visible and near infrared wavebands. The addition of the longer wavelength bands TM5 and TM7 add considerably to the class separation power in this example. Table 6.6(b) shows the effect on the transformed divergence values of adding these two TM bands to the input data. All the values have increased, and in particular there is now good separation of urban from bare rock and scrub and of scrub from other agriculture. The cereals are naturally still confused with other agriculture and the value of $TD_{ij}$ has increased only marginally to 1090 with the addition of the two extra bands.

## Conclusion

The material in this chapter has presented a range of tools with which to analyse satellite remote sensing data. These tools are those most commonly found in turnkey image processing systems: it is not an exhaustive list and there are other versions of these saws, hammers and drills available to the remote sensing scientists. The value of the final product of processing satellite remote sensing data lies neither in the raw data nor in the tools, but in the intelligent combination of the two to fit a particular task. A number of newer techniques such as context analysis have not been included here because their stage of development is not sufficiently mature for inclusion in this introductory text.

The tools discussed in this chapter are widely used in the types of applications which follow in the succeeding chapters. Indeed, the first part of this book has established the nature of the data and the analysis techniques which are then used in Part II of the book, which will now discuss a number of applications of satellite remote sensing.

*PART II*

# Applications of Satellite Remote Sensing

LADY MARGARET HALL LIBRARY
OXFORD

*Chapter 7*

# Land cover and vegetation

## Land cover classification

Perhaps more than any other application of satellite remote sensing the study of vegetation and land use is characterised by classification schemes. There is little or no standardisation in these schemes, and most are designed to fit the needs of a particular project. However, one classification scheme which has received a wide currency in remote sensing is that devised by the US Geological Survey (USGS) for their National Land Use and Land Cover Mapping Program. This programme produces a series of land use and land cover maps for the states of the United States at 1:100,000 and 1:250,000 scales.

The USGS land cover classification is described by Anderson *et al.* (1976) in USGS Professional Paper 964, and is sometimes referred to as the Anderson classification. Levels I and II of the classification are shown in Table 7.1. The classification scheme was designed for applications in the USA, and while it has been used in other parts of the world problems have arisen when applying it in other countries which have landscapes different from those in the USA.

## Agriculture: the Large Area Crop Inventory Experiment

One of the largest experiments in remote sensing of agriculture was the Large Area Crop Inventory Experiment (LACIE) carried out jointly by NASA, NOAA and the US Department of Agriculture (USDA). The objective of LACIE was to develop and test a method of estimating the

TABLE 7.1: US Geological Survey land use and land cover classification system for use with remote sensing data

| Level I | Level II |
|---|---|
| (1) Urban or built-up land | (11) Residential |
| | (12) Commercial and services |
| | (13) Industrial |
| | (14) Transportation, communications and utilities |
| | (15) Industrial and commercial complexes |
| | (16) Mixed urban or built-up land |
| | (17) Other urban or built-up land |
| (2) Agricultural land | (21) Cropland and pasture |
| | (22) Orchards, groves, vineyards, nurseries and ornamental horticultural areas |
| | (23) Confined feeding operations |
| | (24) Other agricultural land |
| (3) Rangeland | (31) Herbaceous rangeland |
| | (32) Shrub–brushland rangeland |
| | (33) Mixed rangeland |
| (4) Forest land | (41) Deciduous forest land |
| | (42) Evergreen forest land |
| | (43) Mixed forest land |
| (5) Water | (51) Streams and canals |
| | (52) Lakes |
| | (53) Reservoirs |
| | (54) Bays and estuaries |
| (6) Wetland | (61) Forested wetland |
| | (62) Non-forested wetlands |
| (7) Barren land | (71) Dry salt flats |
| | (72) Beaches |
| | (73) Sandy areas other than beaches |
| | (74) Bare exposed rock |
| | (75) Strip mines, quarries and gravel pits |
| | (76) Transitional areas |
| | (77) Mixed barren land |
| (8) Tundra | (81) Shrub and brush tundra |
| | (82) Herbaceous tundra |
| | (83) Bare ground tundra |
| | (84) Wet tundra |
| | (85) Mixed tundra |
| (9) Perennial snow or ice | (91) Perennial snowfields |
| | (92) Glaciers |

*Source*: Anderson *et al.* (1976)

production of wheat on a world-wide basis (Erickson 1984). The two principal streams of data in LACIE were:

(1) Landsat MSS data for the estimation of the land area growing wheat; and
(2) meteorological data provided through the World Meteorological Organisation (WMO), US meteorological stations and Noaa satellites for the estimation of the likely yields of wheat.

The combination of the area and yield estimates allowed the computation of the total production of wheat.

LACIE was initiated in 1974, but it had its origins in 1960 when a US group examined the potential role of remote sensing in agriculture, and indeed many of the requirements of agricultural remote sensing led to the specification of the Landsat MSS sensor. The early phases of LACIE were concentrated on the US Great Plains, but the experiment was later expanded to encompass the wheat-producing regions of Canada and the USSR, and some exploratory work was also performed for areas in India, China, Australia, Argentina and Brazil.

## LACIE TECHNIQUES

Estimates of wheat area were made from Landsat MSS digital data. Analysts identified training areas as wheat/non-wheat and computer classification of large sample areas followed based on the training set statistics (see chapter 6). The identification or labelling of wheat/non-wheat was not achieved without difficulty and in general it was not possible to distinguish wheat reliably from other small grain crops, so the identification was one of small grain crops to which historically derived ratios of the proportions of wheat were applied to estimate the wheat areas (Erickson 1984).

Estimates of wheat yield were made using regression models which related past wheat yields to weather statistics in each region. These models used monthly average air temperatures and cumulative precipitation to calculate estimates of wheat yield.

## RESULTS OF LACIE

Probably the most successful and dramatic demonstration of LACIE was in the forecast of the USSR's 1977 total wheat production. Figure 7.1 shows three curves estimating the USSR's production of wheat:

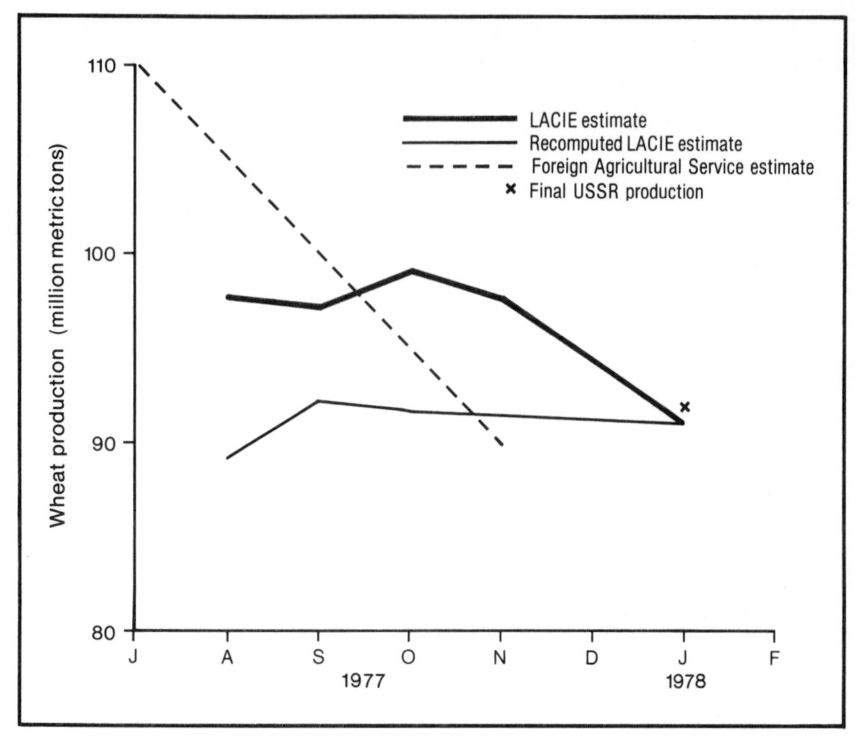


7.1 Estimates of the USSR's total wheat production in 1977. (After Erickson 1984)

(1) The estimates produced by the Foreign Agricultural Service of the USDA based to a large extent on Soviet reports and on reports from US agricultural attachés. These estimates are initially high because of the Soviet goal stated in January 1977 of a wheat crop of 120 million metric tons in 1977.
(2) Estimates by LACIE in 1977 for the final USSR production.
(3) Estimates recomputed by LACIE based on a later, improved method.

The final USSR production of 92.0 million metric tons, published in January 1978, is also shown on Figure 7.1. It is clear that all the LACIE estimates were close to the final figure, and much better than the conventional estimates based on Soviet and other reports. One goal set within LACIE was to estimate production with a 90/90 accuracy, that is to estimate at least 90 per cent of the total production 90 per cent of the time. These results for the USSR do meet this 90/90 criterion.

Similar successes were also reached in the US Great Plains, particularly with winter wheat, and comparison of LACIE estimates with ground data reveal high correlations. Spring wheat was less easy to discriminate than winter wheat because of confusion with other spring small grains, notably spring barley, and because of the narrow strip fields used to grow spring wheat in the USA and Canada. These strip fields are sufficiently narrow to be close to the pixel size of the Landsat MSS, thereby making interpretation more difficult. Similar problems of field size were encountered in the LACIE work in India and China. Clearly the improvements in spatial resolution with the Landsat Thematic Mapper and with the SPOT sensors may well overcome some of these problems and allow more reliable estimation of spring wheat.

### AFTER LACIE

The successor to LACIE was Agristars, developed from the wish of the US Secretary of Agriculture to see a multi-agency programme to develop improved uses of aerospace technology for agricultural purposes. The stated objectives of Agristars, which were much broader than those of LACIE, are given below:

(1) Early warning of environmental or technological changes that may affect the production or the quality of renewable resources.
(2) Improved commodity production forecasts.
(3) Land use classification and measurement.
(4) Renewable resources inventory and measurement.
(5) Land productivity estimates.
(6) Assessment of conservation practices.
(7) Pollution detection and evaluation.

Perhaps because of this broader scope Agristars appears not to have been as successful as LACIE and there has not been a similar depth of application in Agristars.

## Global and regional vegetation analysis

In chapter 5 the characteristics of the Noaa satellites and the Advanced Very High Resolution Radiometer (AVHRR) were discussed. Noaa satellites provide High Resolution Picture Transmission (HRPT), Local Area Coverage (LAC) and Global Area Coverage (GAC) data, and it is the GAC data which have been examined by scientists in the last few years to prepare maps and analyses of continental and global scale vegetation and vegetation change (Yates *et al.* 1986).

Justice *et al.* (1985) present 'the first published global overview of vegetation activity' by analysing the GAC data, and a data set prepared by NOAA from these data, the Global Vegetation Index (GVI) (Tarpley *et al.* 1984). The GVI is produced by calculating the Normalized Vegetation Index (NVI) from the GAC data as shown in equation 6.1 in chapter 6. Each day the NVI is calculated for each GAC pixel (see chapter 5) and the data are resampled to a polar stereographic projection with a 15km resolution at the equator which decreases to 25km at 60° latitude. A weekly GVI image is produced by compositing the highest daily NVI value for each GAC pixel for that week.

Justice *et al.* (1985) used these data as the basis of an analysis of global vegetation phenology for the period April 1982 to January 1983, and took four four-week compositing periods in the northern hemisphere spring, summer, autumn and winter seasons to show the global changes in vegetation seasonally. These seasonal changes are shown in the four maps in the frontispiece. In April–May 1982 the highest GVI values are found in the tropical rain forests and adjacent moist savannas, with high values also in western Europe, eastern China and the eastern USA. By June–July 1982 the green wave of vegetation growth had swept northwards and the agricultural areas of the mid-latitudes had high values as well as the tropical rain forests. In September–October 1982 the mid-latitude areas weakened in their GVI values because of harvesting and vegetation senescence, although the areal extent of vegetation stays much the same. In contrast the southern hemisphere greens up with the onset of spring. By the period December 1982–January 1983 most of the land masses north of 20° N show very low GVI values, with the green wave correspondingly moving southwards in the southern hemisphere.

Justice *et al.* (1985) also show continental and regional scale examples of vegetation phenology using the AVHRR data for Latin America, the whole of Africa (see also Tucker *et al.* 1985), most of the Middle East, western Brazil, the west African Sahel, south-east Asia and India. The value of these global and regional data lies not in the general statements of seasonal vegetation change, which after all are well known, but in the use of the data to make precise statements on where and when vegetation changes are taking place over large areas. This is important not only in understanding global phenology, but also in analysing the wider issues of global biogeochemical cycles and the impact that surface changes can make on global climate (Henderson-Sellers 1984, Badhwar *et al.* 1986).

The Noaa AVHRR data and their derived products have also been used in other regional analyses of land surface change. Hayes and Cracknell (1984) used AVHRR data to prepare a vegetation index map of Scotland;

Schneider *et al.* (1985) used the data to monitor the changing vegetation and water patterns in the Lake Chad basin; Tucker *et al.* (1984) and Nelson and Holben (1986) used AVHRR for assessing deforestation over a 100km × 400km area of Rondonia, Brazil and Richardson (1984) and Tucker and Matson (1985) used the data to analyse the atmospheric and land surface changes brought about by the explosions of the El Chichon volcano in Mexico in March–April 1982. Tucker *et al.* (1986) have compared global NVI data with atmospheric carbon dioxide variations and suggested that the Noaa AVHRR data can be used to estimate terrestrial photosynthesis.

## LIMITATIONS

While there are clearly benefits to be gained from using the Normalised Vegetation Index as an indicator of vegetation phenology, Sellers (1985) has asked the question: what does the vegetation index mean? From his modelling work on the photosynthesis of leaves and vegetation canopies Sellers concluded that the vegetation index is a poor measure of both vegetation biomass and leaf area index when one or more of the following conditions is present:

(1) The leaf area index exceeds the value three.
(2) There are patches of bare ground in the sensor's field of view.
(3) There is an unknown quantity of dead material in the canopy.
(4) The leaf angle distribution in the vegetation canopy is unknown and the solar elevation is high.

The implication of this work is that the vegetation index may be a poor estimator in those areas without vigorous vegetation growth, which are precisely those areas such as the semi-arid grasslands where it is important to have good quality information in order to monitor environmental change.

Sellers (1985) also examined the relationship between the vegetation index and plant biological activity, and found a good relationship between the index and (1) the capacity for photosynthesis and (2) the canopy resistance related to the transpiration of a vegetation canopy. Sellers encouragingly comments that a time integral of the vegetation index should be almost linearly related to the net or gross primary productivity of a vegetation canopy, and concludes (op. cit. page 1368) that 'the simple ratio or vegetation index are poor predictors of leaf area index and biomass should not be surprising, therefore, but the indications that they may be able to provide more profound information regarding photosynthesis and transpiration are encouraging'.

## Regional rangeland surveys

The use of satellite remote sensing in mapping and monitoring rangelands provides what is perhaps the simplest and clearest illustration of the value of satellite imagery. Many rangeland areas of the globe, including deserts and semi-deserts, have only limited environmental information available for them because of problems of accessibility. The visual interpretation of satellite imagery provides a low-cost way of providing some of that environmental information.

In the Middle East a number of studies have shown the value of relatively small-scale rangeland mapping using satellite remote sensing. Mitchell (1981) reports on the use of Landsat images for mapping land systems in Jordan and the Sudan at a scale of 1:1 million, and also uses Landsat imagery of the Middle East and North Africa as input to a Food and Agriculture Organisation (FAO) soil degradation map of the world at 1:5 million scale (see chapter 8). Other applications of satellite remote sensing in the Middle East include those by Krinsley (1974), Allan (1977), Allison (1977), Iranpanah (1977), Parry (1978), Cadet and Desbois (1979), Allan and Richards (1983) and Vass (1983).

### WHITE ORYX PROJECT

Landsat imagery provided valuable environmental information for the White Oryx Project in central Oman (Harris 1983). The aim of the White Oryx Project is the reintroduction of the White or Arabian Oryx (*Oryx Leucoryx*) to the south-eastern Arabian peninsula, from where it is thought to have become extinct in 1972 (Lawton 1978). The project has its base on the Jiddat al Harasis plateau in the central desert of Oman. A small colony of seventeen White Oryx, most of them brought to Oman from Arizona in 1980, began the reintroduction, and these have now been augmented by fifteen offspring.

The Jiddat al Harasis is a flat, stony desert plateau some 200km × 75km in extent situated to the east of the linear dune fields of the Rub al Khali, the Empty Quarter of Saudi Arabia. The vegetation of the area is very sparse and includes scattered distributions of the desert forb and stoloniferous grass with some isolated *Prosopis* and *Acacia* trees. The role of satellite remote sensing in this context lies in the provision of base environmental information for the area to assist with the management of the Oryx. The remote sensing study was conducted in three phases:

(1) Initial satellite image interpretation and map preparation.
(2) Field survey in the Jiddat al Harasis.
(3) Reinterpretation of the satellite images and final map production.

Landsat MSS images for 1972 and 1978 in false colour and in black and white were used at scales ranging from 1:250,000 to 1:1 million. The study area (see Figure 7.2) was covered by four Landsat images and the final maps were produced at 1:1 million scale. The interpretation of the satellite imagery was complemented by field survey, published vegetation surveys and topographic maps. The field survey comprised five transects across parts of the Jiddat al Harasis and surrounding areas, with site descriptions and ground radiometer measurements taken at sample points along the transects. The sample sites are shown on Figure 7.2 and included the principal terrain cover categories of the region, ranging from stony limestone plateaux and pavements, through sand dunes with intervening salt bush vegetation, to shallow wadis with local concentrations of the desert forb and *Acacia* trees.

Figure 7.2 shows the principal location features of the area. The roads and tracks were drawn where visible on the Landsat imagery, and interpolated when there was a degree of confidence about their location. Most of the roads in the area are narrow, graded surfaces of the same material as the surrounding surface lithology, yet they are often visible because of their linearity. In the south-west of the area, around Amal and Shelim, there is clear evidence of the increased oil operations by Petroleum Development (Oman) from 1972 to 1978 shown by the increased size and complexity of the road network.

The terrain cover of the area was interpreted from the Landsat imagery, and the resulting map is shown in Figure 7.3. Ten main terrain cover type categories were identified and these are described in Table 7.2. The emphasis in this classification is on terrain cover type related to vegetation. This is important for determining the likely areas for the Oryx to find food. While the terrain cover classes do not themselves show precisely the area of vegetation because such areas are often so small that they fall below the resolution limit of the MSS, the terrain types are associated with particular types of vegetation. For example, the desert forb is a food plant eaten by the Oryx but not by gazelle or by domestic stock, so the Oryx has little competition for this plant. The desert forb, although sparse, is the most abundant plant species on the Jiddat al Harasis, so the delineation of this area on Figure 7.3 provides an indication of the potential grazing area for the Oryx.

In addition to the map of terrain cover, a map of geomorphological features was prepared from interpretation of the Landsat imagery. This map is shown in Figure 7.4. The emphasis on this map is on linear features: sand

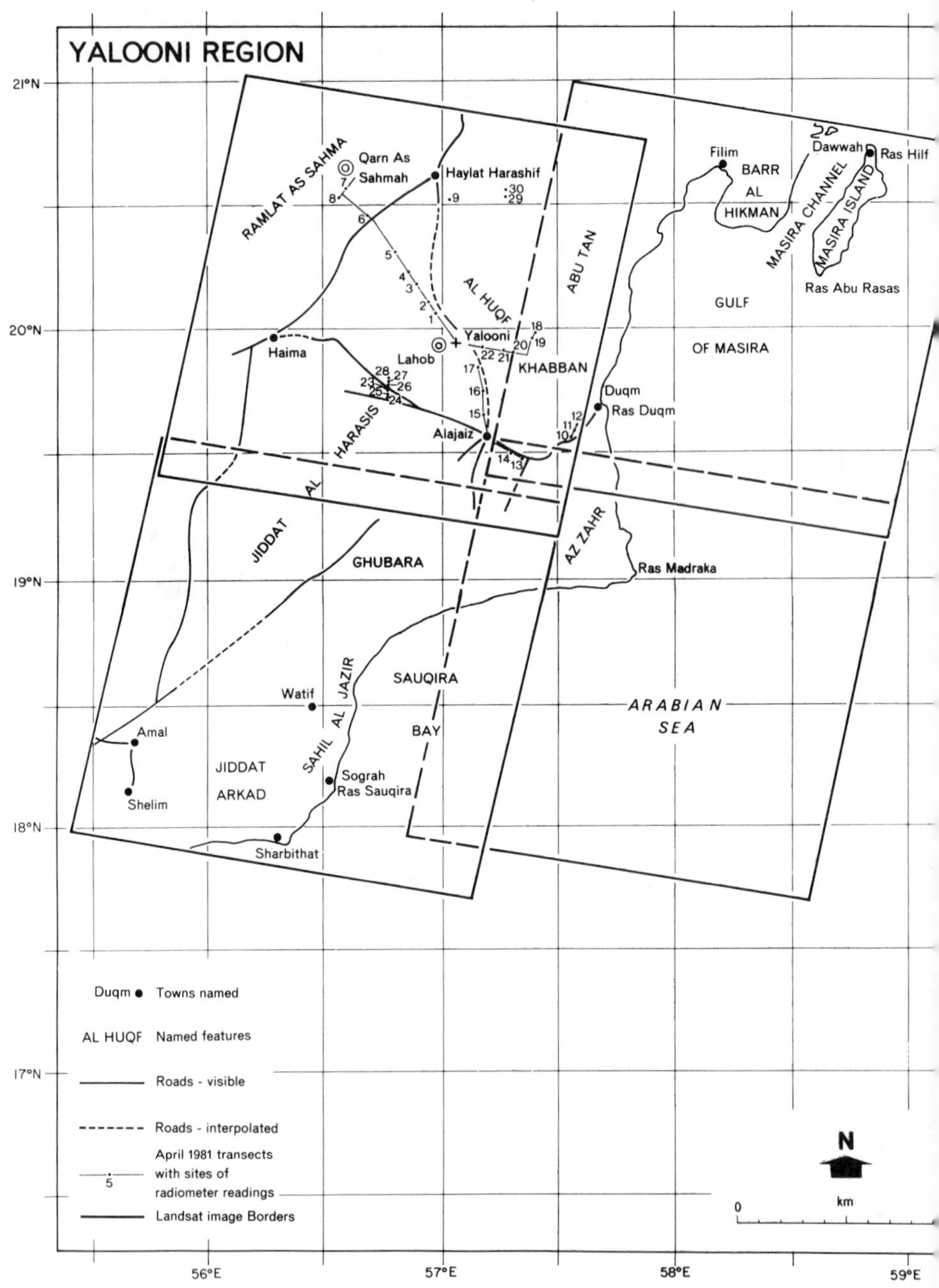


7.2 The coverage of four Landsat MSS scenes in the region of the Jiddat al Harasis, Sultanate of Oman. Location features and the sites of ground data collection points are shown

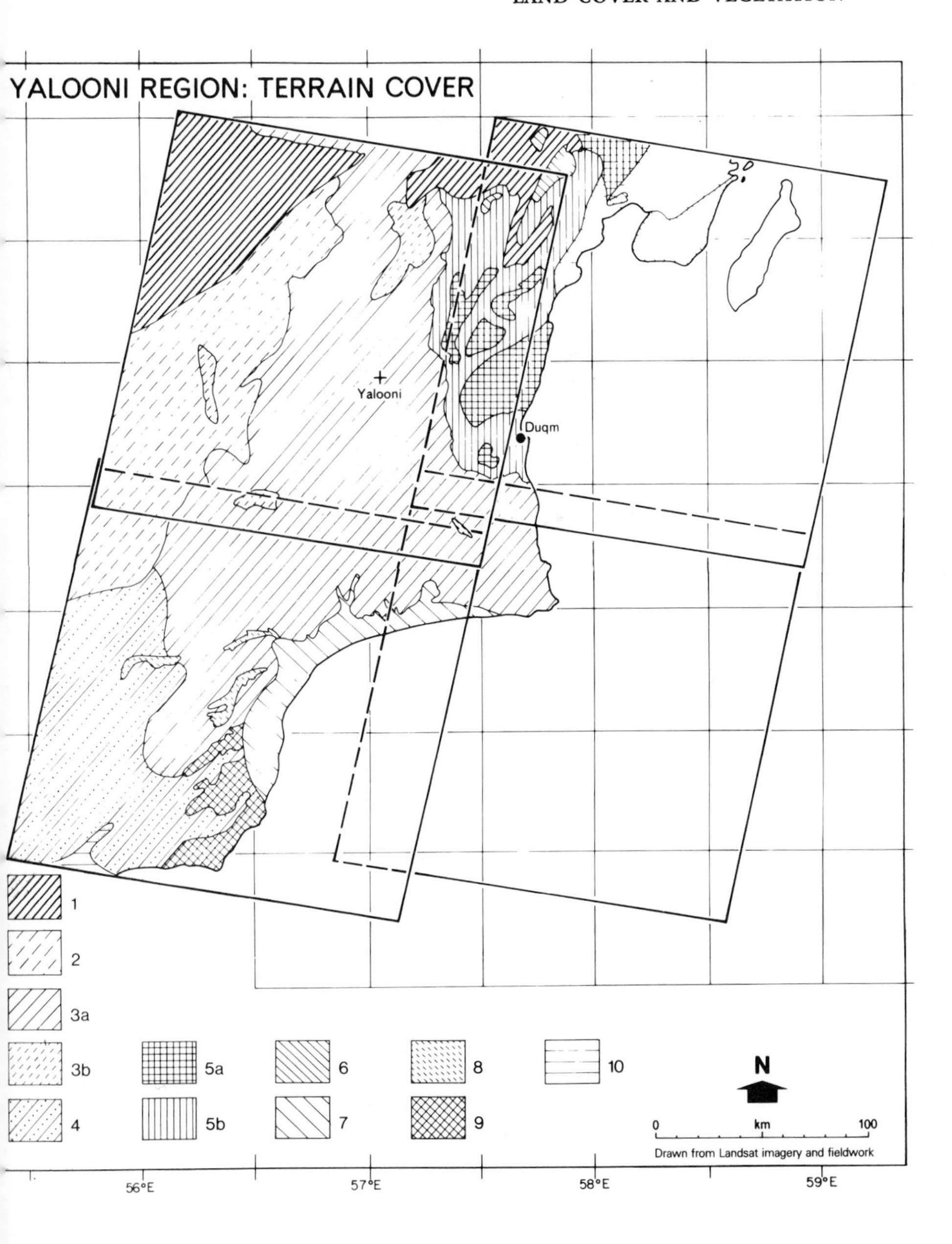


7.3 Terrain cover of the Jiddat al Harasis region prepared from interpretation of 1972 and 1978 Landsat MSS imagery and field survey. The key is given in Table 7.2

*TABLE 7.2: Terrain cover categories identified in the Jiddat al Harasis area, central Oman. This table is the key to Figure 7.3*

| *Category* | | *Description* |
|---|---|---|
| (1) Sand dunes | | Linear dunes c. 15m in height oriented N–S or NE–SW. Stable in location. Some *Prosopis* and *Acacia* trees. |
| (2) Sand plain | | Fine, soft, white sand plain. High albedo. Virtually no vegetation. |
| (3) Jiddat al Harasis | | Stony plateau with stones overlying compacted sand. Sparse vegetation, principally desert forb and small *Acacia* locally enhanced in depressions such as shallow wadis. The small areas of type 3b have less vegetation than type 3a. |
| (4) Southern Jiddat al Harasis | | Similar to type 3 but with increased sand coverage. |
| (5) Eastern plateaux | (a) | Solid geology of limestone, dolomite and shales. Plateaux at c. 70m above wadi floors. No vegetation. |
| | (b) | Intervening sand and gravel deposits. Sparse vegetation. |
| (6) Sebkha | | Very high albedo salt deposits. Flat. No vegetation. |
| (7) Coastal plain | | Wide plain (10–20km) lying between the Arabian Sea and the Jiddat al Harasis. |
| (8) Wadi floors | | Enhanced vegetation in the large southern wadis which are up to 5km wide. |
| (9) Jiddat Arkad | | Higher desert plateau. Sparse vegetation. Deeply incised wadis. |
| (10) Unidentified areas | | |

dunes in the north-west, escarpments in the centre and east and large wadis in the south. The Al Huqf escarpment in the east is a steep, rocky escarpment c. 50m high and separates the higher limestone plateau of the Jiddat al Harasis from the gravel plains to the east.

## RANGELAND INTERPRETATION

This example of Landsat imagery interpretation to provide environmental information for the White Oryx Project shows what can be achieved with satellite remote sensing accompanied by ground survey and other,

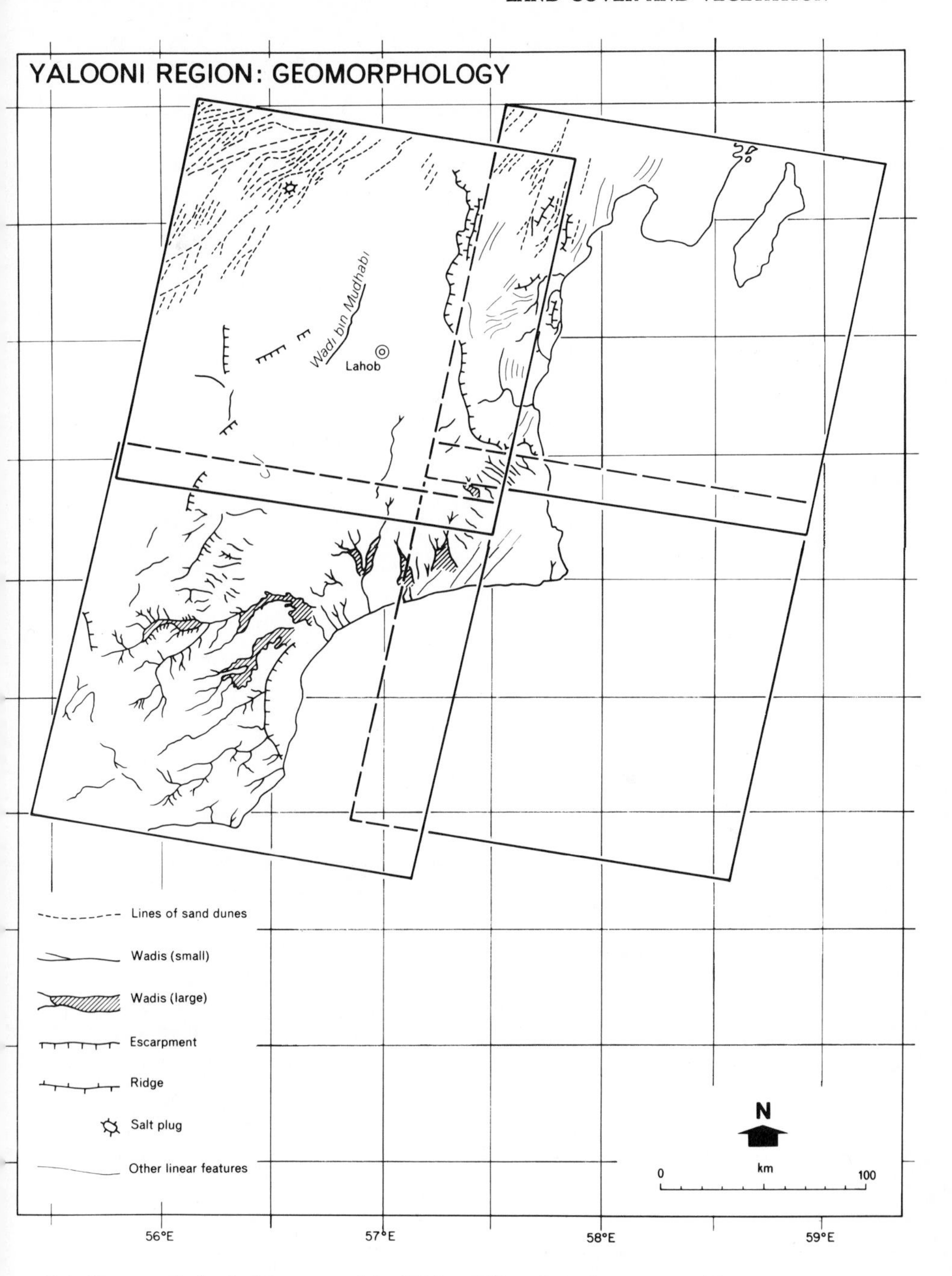


7.4 Geomorphological features of the Jiddat al Harasis region prepared from interpretation of 1972 and 1978 Landsat MSS imagery and field survey

complementary information sources for the study of rangeland resources. For these environments it is often difficult to identify areas of vegetation precisely, but it is possible to see environmental features associated with different vegetation communities. For example, a study in Australia (Löffler and Margules 1980) was able to monitor the area occupied by wombats, not by counting the wombats from space but by analysing their cumulative effect on vegetation.

## Woodlands and forests

### WELSH WOODLANDS

The study of woodlands and forests by satellite remote sensing is aided by the fact that their spatial extent is frequently comparable to the scale of the satellite data, at least for the higher spatial resolution systems. In addition, wooded areas are relatively easy to recognise using satellite imagery (Hardy and Agar 1978, Horne 1984) and the topographic maps of many countries carry information about woodlands and forests.

Bradbury *et al.* (1985) used Landsat MSS and TM data for 27 July 1984 to classify woodland in southern Wales. Their study area comprised stands of spruce, larch and pine on the uplands and oak, ash, beech and elm on the poorer soils and steeper slopes. The other terrain cover types in the area included moorland communities, agricultural areas, bare surfaces and water (see Table 7.3). Data for training and testing a number of classifiers were obtained from 1:25,000 scale Ordnance Survey maps, 1:10,000 scale Forestry Commission stock maps, the Second Land Utilisation Survey and field observations.

Three classification schemes were used in this work: box classifier, centroid classifier and maximum likelihood classifier. The box classifier was found to give a poor classification and so the work was concentrated on the centroid and maximum likelihood classifiers. Table 7.3 shows the percentage accuracy obtained with these two classifiers for each of the terrain categories, although it should be noted that only single figures for accuracy are given by Bradbury *et al.* (1985) and not, as discussed in chapter 6, the accuracy band which should have been stated by the authors.

The two classifiers work well on the broad terrain type classes and both the all coniferous and all woodland classes have accuracies of over 90 per cent using TM data. The individual tree classes show lower performances however, with larch having a highest accuracy of only 42 per cent with TM data and very low accuracy values with MSS data. In general the improved

*TABLE 7.3: Accuracy of terrain type classification of part of southern Wales using Landsat MSS and TM digital data, 27 July 1984. Figures are percentages. An asterisk represents insufficient data*

| | *Classifier* | | | |
|---|---|---|---|---|
| *Terrain type* | *Centroid* | | *Maximum likelihood* | |
| | *Landsat MSS* | *Landsat TM* | *Landsat MSS* | *Landsat TM* |
| Pine | 59 | 80 | 44 | 76 |
| Spruce | 20 | 52 | 36 | 76 |
| Larch | 17 | 37 | 20 | 42 |
| *All coniferous* | 78 | 94 | 81 | 92 |
| Broadleaf | 29 | 56 | 17 | 82 |
| *All woodland* | 75 | 93 | 72 | 96 |
| Scrub | 21 | 10 | 12 | 32 |
| Felled woodland | * | 88 | * | 88 |
| Bracken | 57 | 53 | 42 | 57 |
| Heather | 2 | 35 | 22 | 55 |
| Grass moor | 35 | 45 | 38 | 43 |
| *All moorland* | 61 | 74 | 62 | 41 |
| Pasture | * | 69 | * | 73 |
| Cereals | * | 14 | * | 64 |
| Bare rock | 46 | 1 | 46 | 99 |
| Urban | 78 | 76 | 81 | 69 |
| Mud flats | 10 | 15 | 48 | 72 |
| Marsh | 1 | 5 | 27 | 47 |
| Open water | 94 | 98 | 85 | 98 |

*Source*: Bradbury *et al.* (1985)

spatial resolution of the TM compared with the MSS also improves the accuracy of classification for all the classes, with a few interesting exceptions (see also Irons *et al.* 1985).

The urban class shows deterioration from MSS to TM data because the varied nature of urban areas, which include grassland, parks, woodland and lakes as well as tarmac, concrete and brick, produced misclassifications with higher spatial resolution data. The bare rock classification is improved using TM data with a maximum likelihood classifier, but deteriorates from MSS to TM data using the centroid classifier: this is probably an effect of the classification algorithm rather than a relationship between the bare rock terrain type and improved spatial resolution.

Bradbury *et al.* (1985) conclude that mid-summer Landsat TM data can be used to map woodland in general and some woodland species in particular with high accuracy, and that the accuracy values are likely to increase if multitemporal data are used and if digital topographic data are available to include the effects of slope and aspect on reflectance values.

## CALIFORNIAN FORESTS

### *Klamath*

The point which Bradbury *et al.* (1985) note about the importance of topography is a common problem for the analysis of satellite remote sensing data of woodland and forest areas, which are frequently found in hilly or mountainous terrain. Topography has the effect of increasing reflectance on those slopes facing towards the sun and decreasing reflectance on those slopes facing away from the sun, even though the woodland may be identical on both slopes. Strahler (1981) acknowledges this problem, and in a study of the Klamath National Forest in northern California includes digital topographic data in a classification using Landsat MSS data for forestry analysis.

Strahler used an automated version of classical photointerpretation, namely the analysis of tone, texture and terrain. The tone data were provided by the four bands of Landsat MSS data for 14 July 1976, the texture data by calculating the standard deviation of each 3 × 3 pixel window of the MSS band 5 data and the terrain data by elevation data on a 50m grid from the US National Cartographic Information Centre. From these sources eight data planes were co-registered: the four Landsat MSS bands, the texture channel, the digital elevation data and the digital slope and aspect derived from the elevation data. Strahler took a simple approach to the inclusion of the topographic data by determining shadowed and non-shadowed regions. On an initial analysis of the MSS data it appeared that the areas of moderately open to sparsely stocked forest stands on slopes facing away from the sun were being confused with densely stocked stands on slopes facing towards the sun. Strahler took the digital elevation data and for each pixel calculated z, the angle between a normal to the land surface and the sun at the time of the Landsat overpass. Where z was 60 ° or greater then a shadowed pixel was declared, and so for the whole co-registered data set shadowed and non-shadowed regions could be identified.

Having performed this stratification, the Landsat MSS data together with the texture data were cluster classified to produce a large number of classes (over a hundred). These classes were edited to combine some similar classes, then a final classification was performed using a combination of a box classifier and a maximum likelihood classifier. Elevation and aspect data were also used to aid the classification as red fir, mixed conifers and ponderosa pine were known to occupy different elevations. The classified data could then be used for assessing timber volume as well as producing regional forest maps. From an analysis of 114 timber volume samples Strahler found that the use of Landsat MSS data gave results comparable at a statistically significant level to conventional photointerpretation techniques.

### *El Dorado*

A second example of Californian forestry analysis is provided by the work of Logan (1983) on the El Dorado National Forest. The aim of Logan's work was to estimate coniferous forest biomass over large areas by using a nested sampling approach which involved three stages:

(1) Ground measurement of biomass.
(2) Calibration of a Landsat-based forest stratum map with biomass ground truth data.
(3) Extension of the relationships to Noaa AVHRR data for large areas.

Logan found that for forest biomass the various forms of vegetation index applied to Noaa AVHRR data gave poor estimates of biomass because of non-linearity in the red/near infrared feature space. This non-linearity is shown in Figure 7.5 which shows contours of coniferous forest biomass plotted in the feature space defined by AVHRR channels 1 and 2.

Logan found that the best method of forest biomass estimation was a mixed box and maximum likelihood classification of the AVHRR data which allowed for the non-linearity problem. Using this technique the correlation between the biomass predicted by the classification method and the ground-based biomass data was 0.85, which for satellite remote sensing is a respectable result.

## REQUIREMENTS AND REGIONAL EXAMPLES

The examples of satellite remote sensing for woodland and forestry discussed above illustrate the value of satellite data for classifying such areas. Hildebrandt (1983) suggests though that more should be expected from satellite remote sensing of forestry, including:

Forest cover monitoring.
Forest resources inventory.

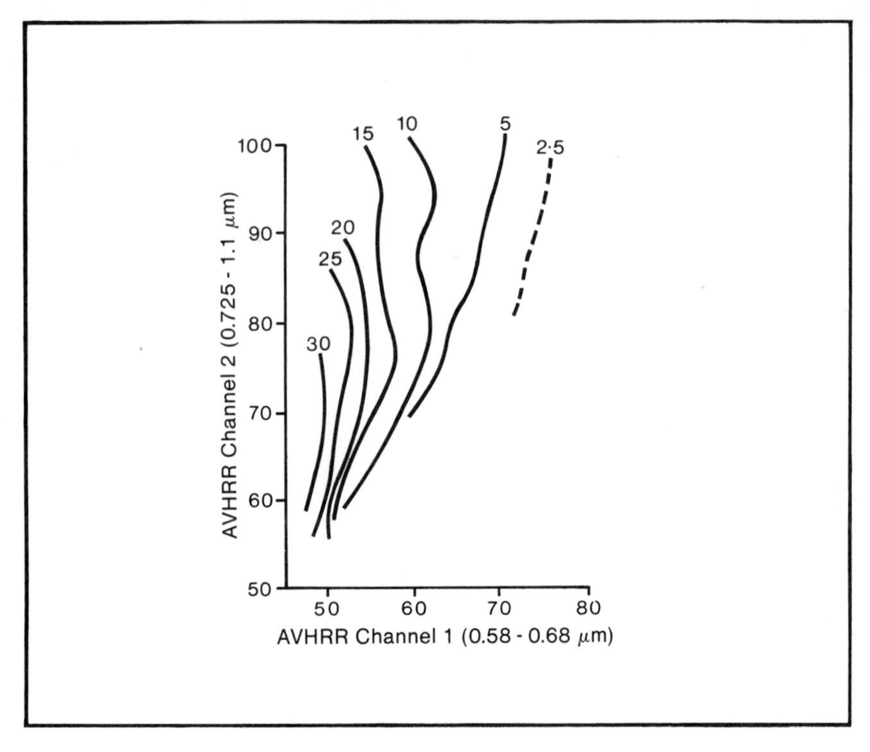


7.5 Coniferous forest biomass contours in million kg drawn in the Noaa AVHRR feature space for the El Dorado National Forest, California. (After Logan 1983)

Classification and mapping of forest types, age, density and management units.
Monitoring changes of ecological, silvicultural and economic conditions of forests.
Damage and disease detection and assessment.
Forest fire detection and control.

The requirements are easy to list but rather more difficult to put into operation given the wide global variety of forest types, the limited information available on many tropical and boreal forests and the range of satellite remote sensing data needed to accomplish the tasks. However, Hildebrandt does quote some examples which illustrate what can be achieved.

The Ontario Centre for Remote Sensing (OCRS) in Canada has used Landsat data for forest inventory and mapping of Ontario north of 52° N, which covers an area of more than 200,000km$^2$. The OCRS has produced terrain maps at scales from 1:50,000 to 1:250,000 on a UTM projection which show classifications of black spruce, jack pine, poplar and white birch forest areas, and water, wetland and forest areas. Horler and Ahern (1986) give a detailed example of forest mapping in western Ontario using Landsat TM data.

In Alaska Landsat data have been used to provide an initial forest classification which is then further refined using medium- and large-scale colour infrared aerial photography for selected sample points. In Brazil Landsat MSS and RBV images have been interpreted visually to assess deforestation and reforestation in Amazonia; in west Africa Landsat imagery has been used in forest reconnaissance level surveys and in the USSR Meteor and Cosmos satellite imagery is used to assist in forest fire prevention and control, forest cutting operations, and reforestation schemes.

## Urban areas

### URBAN REFLECTANCE

Urban areas are often strikingly visible on many forms of satellite remote sensing imagery. DMSP visible wavelength imagery taken at night shows the cities of Europe and North America standing out as bright white spots, and the imagery looks almost like a map of the distribution of cities. On Landsat TM data the internal structure of major cities such as London, Paris and New York is clearly visible. However, although urban areas are often strikingly present on such images, on other occasions they are difficult to discern and do not present a consistent spectral response. As discussed earlier in this book (see chapter 2), the spectral reflectance curves for bare surfaces show similar values in the visible and near infrared so that unlike vegetation there is no simple contrast between these two wavelengths. On plots of near infrared to visible reflectance urban surface types are scattered along the plane of soils (see Figure 7.6) and can be confused with a variety of other surface types. For example, the city of Omdurman (population over 300,000) in the Sudan is virtually invisible on Landsat MSS images of the area as the reflectance of these urban surfaces is similar to that of the surrounding bare rock and alluvial deposits.

### LANDSAT DATA

The US Bureau of the Census has used Landsat imagery in the mapping of the US Standard Metropolitan Statistical Areas (SMSAs), and began the

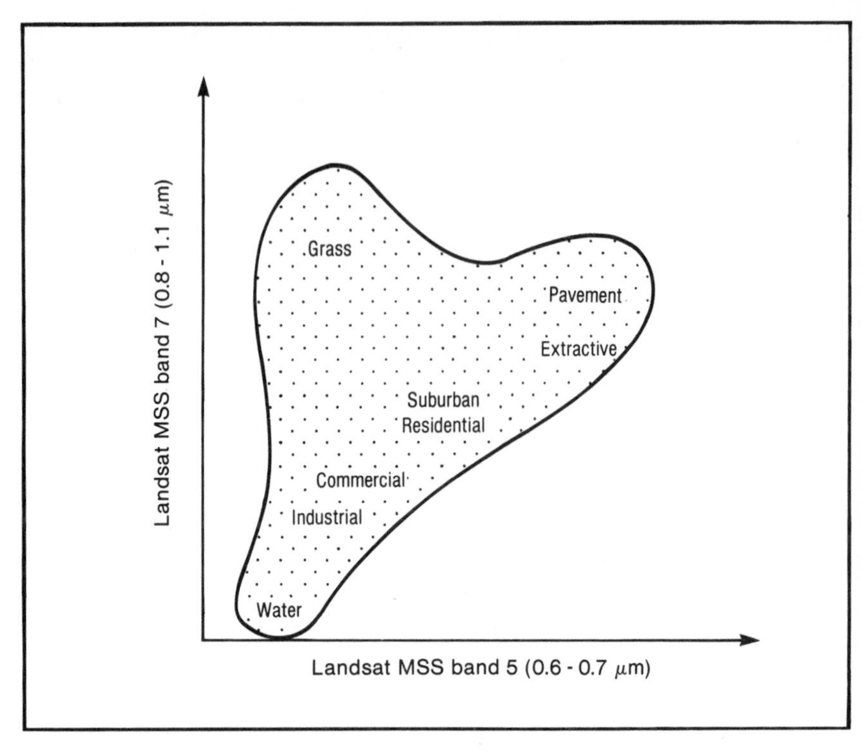


7.6 Partition of the Landsat MSS band 5 and band 7 feature space for analysing urban areas. (After Jensen 1983)

work by examining five selected SMSAs: Richmond, Seattle, Austin, Boston and Orlando (Jensen 1983). Enhanced Landsat MSS images were visually interpreted and the digital data classified by partitioning the MSS5/MSS7 feature space into arbitrarily defined regions. Figure 7.6 shows the types of urban areas and urban functions associated with this feature space partition. The diagram shows the plane of soils along which the urban surfaces lie and the consequent potential for confusion with other bare surfaces. A comparison of large-scale aerial photography with enhanced Landsat images of the SMSAs showed that in both data sets the outer edge of the SMSA was identifiable, but that large individual structures, such as shopping centres could not be identified on the Landsat imagery.

In the UK the Department of the Environment (DoE) investigated the use of Landsat MSS digital data to assist with the monitoring of urban growth (Carter 1979, Carter and Stow 1979). Landsat MSS data for five towns were used: Reading, Northampton, Preston, Hemel Hempstead and Newmarket. The ground checking data were provided by 1:60,000 scale aerial photography. The Landsat data were first transformed by principal components analysis and then classified using a maximum likelihood classifier. The results showed that urban areas were correctly classified with an accuracy typically in the range 70–80 per cent and the rural areas in the range 80–95 per cent. While these results are generally comparable with other classifications of Landsat MSS data they are inadequate for the DoE's monitoring purposes. The work reported by Carter for the Reading area shows that of the 202ha of actual growth between 1973 and 1976, 176ha were correctly classified, but a further 1357ha were also classified as areas of urban growth. The growth of urban areas in England and Wales is typically of the order of 1 per cent per annum, so these results from analysis of Landsat data are clearly insensitive to this order of change.

Carter extended the analysis by building elimination rules into the study. In this second stage only areas of growth above 2.5ha and lying within 200m of the existing urban areas were accepted for classification as urban growth. This method produced a cleaning up of the classification map and the elimination of many of the wrongly classified pixels, although it is not clear whether this is sufficient to make the analysis sensitive enough for the DoE's needs.

If urban areas can be identified on satellite remote sensing imagery then change through time can be estimated by analysing sequential images (Forster 1985). Figure 7.7 shows a map of urban development in Kuwait City from 1973 to 1980, and was produced from visual interpretation of Landsat MSS imagery for 1973 and 1978 and an RBV image for 1980. The new developments in the southern borders of the city are shown in stippled shading and the new roads in the south, which comprise part of the ring road system, are shown as thicker lines. The key to this interpretation was the structure of the road network rather than simply spectral reflectance as there was confusion in computer classification of Landsat MSS data of the urban area with the coastal mud flats and some of the desert surfaces.

## RADAR DATA

The limited amount of data available from Seasat and the Shuttle Imaging Radars show that urban areas are identifiable by their high backscatter. The structures in urban areas act rather like corner reflectors and so return a high proportion of the radar energy back to the sensor. Figure 3.5 in chapter 3 shows the high backscatter of settlements on a SIR-A image of the Sousse region of Tunisia. Bryan (1983) analysed L-band radar imagery of Los Angeles and concluded that internal city structure could be determined because the different orientations and spacings of roads produced different effects on the radar images. In many cases radar data can

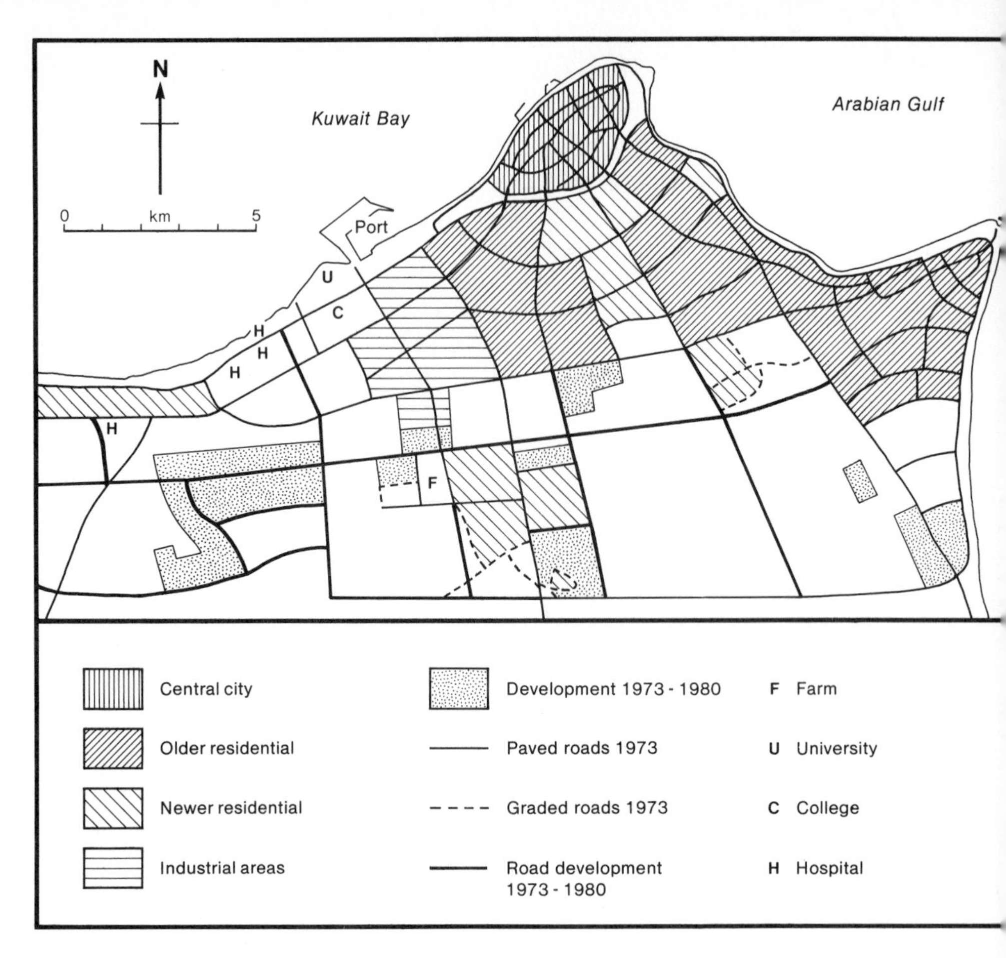


7.7 Change in Kuwait City mapped from Landsat MSS images for 1973 and 1978 and a Landsat RBV image for 1980. The features identified by the letter code were taken from contemporary map information

act as a useful complement to Landsat data, and when co-registered better discrimination can be achieved than by using Landsat data alone because the roughness information provided by radar provides an extra dimension to the data base not covered by visible and near infrared remote sensing (Toll 1985).

## Cartographic considerations

If satellite remote sensing is to become operational then it is important to determine the mapping standards to which the data comply. In April 1981 a Remote Sensing Society workshop was held in London to discuss the role of satellite imagery for topographic mapping (Dowman 1981). The conclusions of the workshop were as follows:

(1) The cartographic use of Landsat MSS data is limited to 1:500,000 scale mapping with no height information.
(2) There is no certainty that planned satellites will be able to produce data sufficient for economical mapping at scales greater than 1:100,000 with 50m contours.
(3) There is a requirement for satellite imagery to provide maps of the unmapped areas of the world at 1:50,000 scale with 20m contours.
(4) Radar has considerable potential.

This rather conservative view, due in no small measure to the conservative stance of the Ordnance Survey concerning satellite remote sensing, has had to be modified over the succeeding years, and at another Remote Sensing Society conference in London in September 1985 it was clear that Landsat TM data were of a high quality for cartographic applications. At that conference Bryant *et al.* (1985) drew the following conclusions:

(1) Landsat TM digital products meet or exceed 1:100,000 scale US map accuracy standards for horizontal control, and may even meet 1:50,000 scale map accuracy standards (Welch *et al.* 1985).
(2) Data corrected to the Universal Transverse Mercator (UTM) projection are preferable to those on the Space Oblique Mercator (SOM) projection because with UTM the offset between ground coordinates and satellite coordinates is constant throughout the scene so only a single rectification is necessary.
(3) Landsat TM data should not be considered for elevation mapping purposes as the amount of horizontal displacement between adjacent paths where scene overlap occurs is not enough to measure relief and obtain relative or absolute elevation contours.
(4) Relief has virtually no impact on horizontal map accuracy statistics in all terrain types except those with very large relief displacements over short distances.

The encouraging results reported at the 1985 conference suggested then that Landsat TM data were suitable for at least 1:100,000 scale mapping and that SPOT data may well be suitable for 1:50,000 scale mapping with 40m contour intervals (see also Welch 1985): with the launch of SPOT in 1986 this has turned out to be correct (Jaques and Lopez 1986).

Togliatti (1986) reported on the early use of Large Format Camera images for space cartography. Because of the forward motion compensation built into the camera system a long exposure time was possible and so a high spatial resolution of c. 5m was achieved in areas of high scene contrast. Tests in the Po Valley in northern Italy with control points and tie points taken from existing maps, and using photogrammetric adjustments to the images, showed displacement errors of only 5m in both horizontal and vertical dimensions.

*Chapter 8*

# Geology and soils

Satellite remote sensing in geology represents perhaps the most common application of satellite remote sensing data in the environmental sciences. Geologists have used aerial photography extensively, and termed this photogeology, and for many applied uses the extension to satellite data is simply a way of raising the height of the data gathering platform. Many of the visual interpretation techniques of photogeology can be applied to satellite remote sensing data, but extended spectrally to include new parts of the electromagnetic spectrum such as the middle infrared and the thermal infrared. An important type of information available to the geologist from satellite remote sensing is the structural information revealed by lineaments on the imagery (Short 1982). Radar imagery in particular has been useful in structural studies because radar illuminates the terrain obliquely and in areas of faults and folds the images can often clearly illustrate structure by the shadow effect. A second role of satellite remote sensing in geology is in the identification of rock type at the surface using image analysis techniques such as those described in chapter 6. This rock type information is useful in geological mapping in areas of little or no vegetation, although some geologists have suggested that the structural information is of greater importance to understanding the geology of an area than is the surface rock type.

## Lithology and rock type

### DISCRIMINATION AND IDENTIFICATION

Two main elements have to be considered for the analysis of rock type: discrimination and identification. Discrimination provides the distinction between surface lithologies based on their image properties such as tone, colour and/or texture, and allows the geologist to prepare a map of the surface characteristics of a study area. Identification is the important step of deciding which rock types correspond to the units discriminated at the previous stage. The discrimination stage can be performed relatively easily by visual interpretation or digital analysis, but the identification stage needs extra information on the spectral characteristics of rocks.

Information on the spectral response curves of rock types has been obtained from three sources: laboratory spectra, field radiometry spectra and spectra from airborne or spaceborne sensors. Hunt and Salisbury in a series of papers in the journal *Modern Geology* have collected laboratory spectra for over 150 rock types (see for example Hunt and Salisbury 1970, 1976, Hunt *et al.* 1973) and these have formed the basis of much subsequent comparative work using field and satellite data. Unfortunately, the spectra of most rock types do not have as distinctive a curve as that for vegetation, but a number of general relationships do exist (Williams 1983). Figure 8.1 shows the spectral response curves of four surfaces collected using a field reflectance spectrometer which illustrate some of the characteristics discussed here. The wavebands of the Landsat TM are also shown.

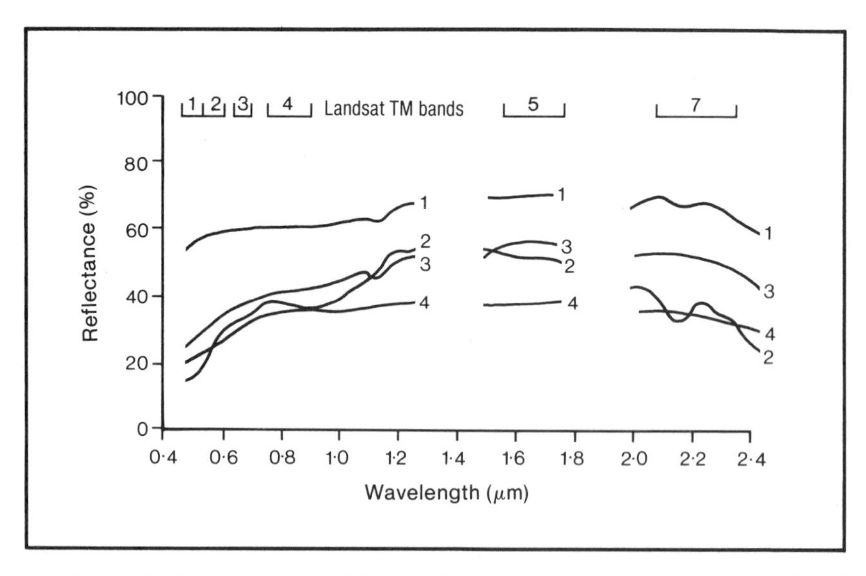


8.1 Spectral reflectance curves of four surface types: (1) volcanic rocks, (2) altered latite, (3) limonitic volcanics, (4) tan soil or rhyolite. The data were collected with a field reflectance spectrometer. (After Williams 1983)

(1) Limonitic rocks (i.e. those with iron oxides present) show a characteristic depression in their spectral response at wavelengths below 0.5 μm. This is caused by an intense iron absorption band at short wavelengths. A second shallow depression is commonly found between 0.85 and 0.95 μm.
(2) Carbonates and clay minerals have distinctive absorption bands in the 2.1–2.5 μm region, although these are normally narrow features in the spectral response curve and can only be determined accurately with high spectral resolution sensors. Hydrothermally altered rocks commonly contain clay minerals (e.g. kaolinite and alunite) which have this diagnostic feature, and one reason for the inclusion on the Landsat TM of the middle infrared band 7 (2.08–2.35 μm) was the value of this spectral region for identifying areas of hydrothermally altered rocks.
(3) An increase in silica content shifts the absorption in the 8–14 μm thermal infrared region to shorter wavelengths, so an increase in silica can be detected if the wavelength of the absorption maximum decreases at thermal infrared wavelengths.

Other conclusions have been reached from the analysis of laboratory spectra of rocks, but for satellite remote sensing the problems are threefold. Firstly, the weathered surface may have a different spectral response curve than the laboratory spectrum of a crushed version of the same rock type. Secondly, the scale of satellite remote sensing pixels commonly involves spatial variation of surface type within a pixel and, thirdly, the relatively broad wavebands used in most remote sensing satellites are not very sensitive to narrow band absorption features. New, experimental sensors flown on aircraft have tackled this problem by using very narrow spectral bands for geological analysis (Weber 1985).

## BAND RATIOS AND PRINCIPAL COMPONENTS

Band ratios have been used in geology as they have been in vegetation studies. Band ratios have been used for the production of colour composite images of Landsat data by calculating the ratio images of MSS 4/5, 5/6 and 6/7 and then displaying each of the new ratio images as blue, green and red respectively. This is commonly a useful enhancement of the data which aids visual interpretation, particularly of arid zone imagery.

Ratios of 1.6/0.5 μm wavebands help to define limonitic rocks: a high value of this ratio indicates iron absorbing limonitic rocks and a low value non-limonitic rocks. A 1.6/2.2 μm ratio helps distinguish hydrothermally altered rocks as such rocks have high values in this ratio. The addition of (say) a 0.8/0.65 μm ratio will distinguish healthy vegetation, and a colour composite of blue, red, green for the three ratios included in this paragraph will help identify rock type and lithological differences.

Rothery and Milton (1981) report work on attempts to identify ophiolites, which are preserved remnants of a Cretaceous oceanic lithosphere, in the Jebel Akhdar mountains of Oman using Landsat MSS and ground radiometer data. They found that the lithologies in the ophiolite were not readily distinguishable on the basis of their individual reflectance properties, but they did calculate ratios of Landsat MSS bands 4/5, 5/6 and 6/7. The ratios showed only slight improvement in discriminating between different lithologies of the ophiolite, except for the MSS 4/5 ratio which was useful in identifying gossans, areas of hydrothermally altered rocks normally associated with copper mineralisation.

Baker and Baldwin (1981) also used the MSS 4/5 ratio to identify gossans in the El Salvador region of northern Chile, part of the porphyry copper belt. They found that band ratios provided some useful information, but their preferred technique was principal components analysis. Baker and Baldwin used this technique to maximise the differences between the spectral response of different lithologies, and by using the first three principal component images were able to classify gossans. All known gossans greater than c. 150m in diameter were identified correctly, several unknown gossans were identified, and only one classified gossan proved to be false.

One problem with principal components analysis is that it is scene-specific. The technique provides a useful way of enhancing multispectral data, but the quantitative information concerning the components learned in one study may not readily be applied in a different study area without a new principal components analysis being performed.

## GEOBOTANY

As much of the land surface of the Earth is covered by vegetation then often the surface geology cannot be sensed directly. An indirect measure of the geology can be obtained by analysis of the vegetation cover, and this can be particularly successful where the local rock type has a distinct effect upon the vegetation (Morrisely *et al.* 1984). Sabins (1978) reports work by the US Geological Survey in the state of Maine where the spectral reflectance curves of Balsam Fir and Red Spruce were measured over the wavelength range 0.4–1.1 μm for trees growing in normal soils and in soils enriched with copper and molybdenum from the underlying solid geology. The spectral reflectance curves are shown in Figure 8.2 and from these it is clear that the mineralisation increases the visible reflectance of both tree species, but that contrasting effects occur in the near infrared. This work was performed using a field spectrometer (Yost and Wenderoth 1971) and Sabins (1978) comments that these specific studies may be difficult to apply to Landsat MSS data, and indeed other satellite data, because of the spatial integration effect of large pixels. However, it is clear from Figure 8.2 that effects are produced by the underlying geology and that these may be detectable under suitable conditions.

Labovitz *et al.* (1985) used radiometer measurements of leaves of oak and tulip poplar trees to examine the effects of mineralisation of copper, iron, lead and zinc. They chose mineralised and non-mineralised areas around the town of Mineral, Virginia, as representative of a larger area running from New Jersey to Georgia, and measured the spectral reflectance of leaves using a radiometer equipped with bands equivalent to Landsat TM3, TM4 and TM5 (see chapter 5). They found that reflectance is higher for leaves growing over mineralised soils, and that the effect is greatest in the late summer and early autumn. Again, this work is based on small area measurements and the results may not be directly extendable to pixel sizes of 20–80m.

Lulla (1985) and Weber (1985) give examples of geobotanical remote sensing, particularly related to mineral exploration, and Lulla includes a bibliography of 213 items relevant to the subject.

## Lineaments and structure

The surface expression of folds, faults, joints and many other geological features is often in the form of lineaments. On Landsat imagery taken in the winter season when the solar elevation angle is low, and on radar imagery with an oblique viewing geometry linear features can frequently be observed. Geologists have used such satellite imagery to detect the presence of lineaments and to measure their frequency, orientation and length. This information can then be related to geological structure when used in conjunction with conventional geological information and has often resulted in extensions to structural information in previously mapped areas.

For the purpose of lineament analysis, Landsat data can be edge enhanced by passing over the image a small (typically 3 × 3 pixel) gradient filter which enhances the sharpness of the image. The lineaments can then be detected visually and analysed. Mamula and Voight (1982) give examples for Iceland. Gonzalez (1982) has expressed considerable confidence in this

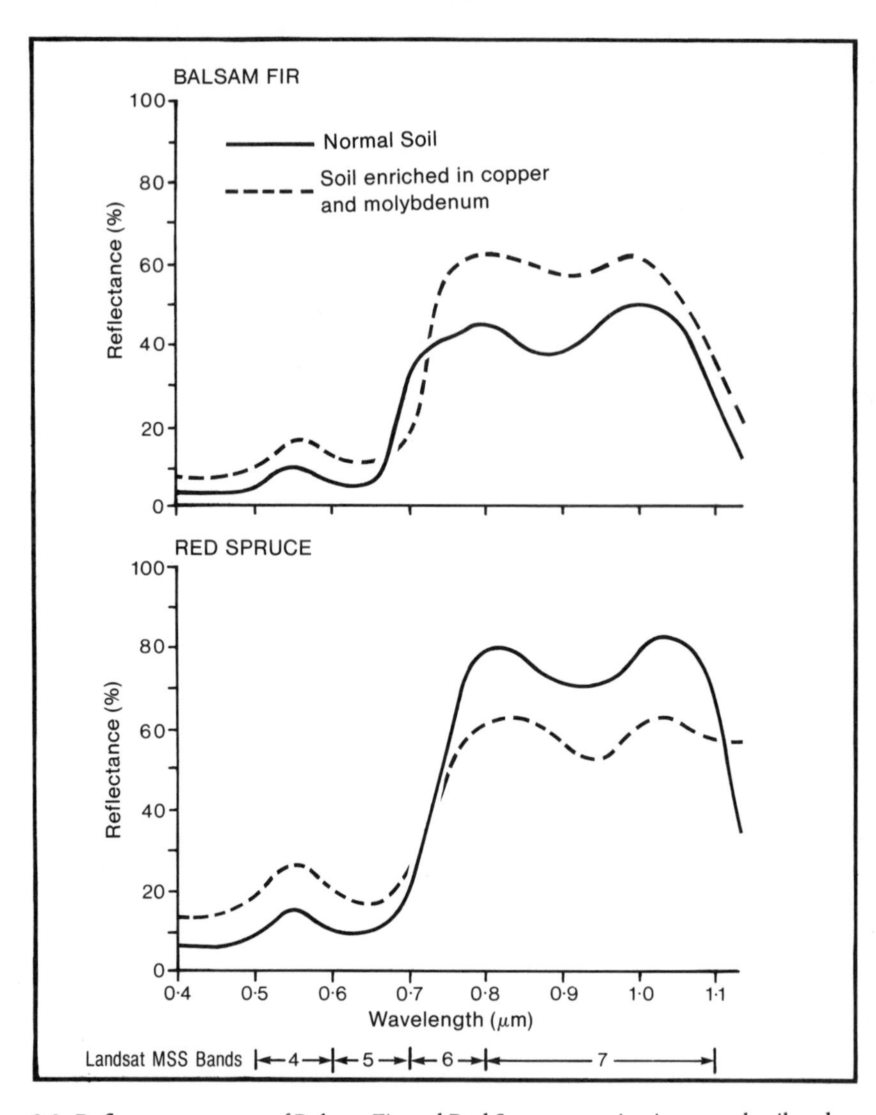


8.2 Reflectance spectra of Balsam Fir and Red Spruce growing in normal soil and soil enriched with copper and molybdenum. (After Sabins 1978)

form of analysis for hydrocarbon exploration and suggests that lineament analysis using Landsat imagery 'is probably the most beneficial technique used in "phase-one" hydrocarbon programs'. He cites examples from the Appalachians, Montana, Kansas, Kenya and the USSR.

## WESSEX BASIN

Lake *et al.* (1984) used Landsat MSS and TM imagery of southern England to map the lineaments in the Wessex basin, and their map drawn from TM data is shown in Figure 8.3. They mapped both clear and subtle lineaments, screened the data to remove lineaments related to cultural features such as roads and railways, collated the data in rose diagrams, and summarised the principal lineament trends. Five major trends were recognised and related to elements of the Variscan and Caledonian geological structures of the area, and conclusions were drawn on the value of the structural information for oil exploration.

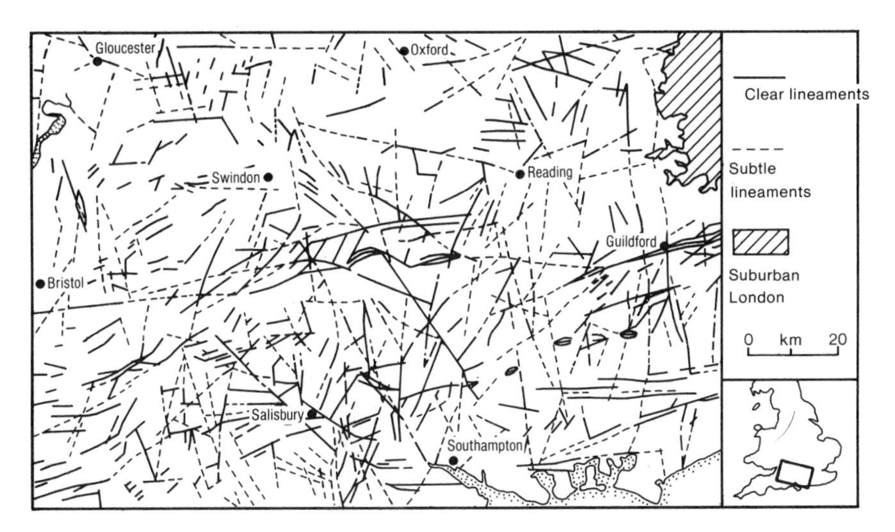


8.3 Lineament map of the Wessex basin compiled from interpretation of a Landsat TM band 5 image, 4 February 1983. (After Lake *et al.* 1984)

A problem which Lake *et al.* (1984) note, and which is common to this type of analysis, is that NW–SE trending lineaments are commonly missed because at the local data capture time for Landsat (c. 10.00 hours) the sun is shining from the south-east in the northern hemisphere mid-latitudes, along the trend of the NW–SE linear features and so does not highlight them, but preferentially highlights the SW–NE trends. This problem can be accommodated by using other satellite data, for example Seasat, which have a different imaging geometry.

## CRETE

Stefouli and Osmaston (1984) applied the techniques of lineament analysis in a geological study of Crete. They note that Landsat linear features belong to three main geologic classes, viz. fractures (faults and joints), lithological contacts, and folds. These authors go further than most in attempting to identify the relationships between the characteristics of linear features observed on a Landsat image and their geological nature. The relationships are listed in Table 8.1, although the authors do point out that extra information on structure and lithology is needed for a correct analysis. Four main systems are identified in Crete:

(1) A system of old fault lines and traces of folding of a general E–W direction.
(2) A system of nearly N–S and E–W faults and fractures related to recent tectonic activity.
(3) A system of SE running faults resulting from recent tectonic activity.
(4) A system of NE running fractures which only develop on the Pindos and Tripolitza nappes and are not detectable in the field.

*TABLE 8.1: Relationships between the characteristics of linear features observed on Landsat MSS images and their geological nature*

| *Characteristic of linear feature* | *Geological inference* |
|---|---|
| Boundaries between different image patterns | Fault |
| Texture features | Fractures |
| Relief change features | Major fault lines |
| Boundaries between different image textures | Lithological boundaries, or faults which bring into contact different lithologies |
| Tonal features | Lithological boundaries or faults |
| Light lines | Fractures |
| Dark lines | Minor fractures |

*Source*: After Stefouli and Osmaston (1984)

# Exploration geology: examples

By a combination of rock type discrimination and lineament and structural analysis, satellite data at small scales (typically 1:250,000 to 1:1 million) have been useful in mineral exploration. Williams (1983) gives examples of the role of remote sensing in economic geology, including mineral exploration, petroleum exploration, coal exploration and geothermal exploration. Satellite remote sensing imagery is often of great value in such studies by providing images of large areas which reveal features previously unknown when only field observations or large-scale aerial photography are used. Radar imagery has been of particular value recently because SIR-A has shown the ability of L-band synthetic aperture radar to penetrate sand surfaces in hyper-arid areas (Ford *et al.* 1982). Much new geological information about the sub-surface geological composition some 1–3m below the sand surface has been discovered in parts of the Sahara desert from Algeria to Egypt. In addition, on the SIR-A images massive shear zones were identified in Africa and China, volcanoes in Turkey and Indonesia, fold mountains in Iran and intrusive granites in Australia. Figure 8.4 shows a SIR-A image of the southern part of the Wahiba Sands, Sultanate of Oman, and reveals the structure of the massive sand dunes running north–south, and backscatter variations related to subsurface features.

## SOUTH-WEST ENGLAND

Moore and Camm (1982) used Landsat MSS data to assist in prospecting for tin and tungsten in south-west England. They enhanced Landsat digital data of the area by principal component transformation and directional enhancements, and were able to map tectonic features and rock type distributions in the study area at 1:250,000 scale. Figure 8.5 shows the positions of the bedding trace and fracture lines interpreted from the imagery together with other geological and geophysical data for the area. The enhanced Landsat images were able to assist the geologists in preparing or redrawing maps of those areas affected by thermal metamorphism, potential sites for fault-controlled mineralisation, concealed small granite bodies and kaolin alteration. These results were achieved in an agricultural area with almost no bed-rock outcrops and no undisturbed surface signs of mineralisation.

8.4 SIR-A image of the southern part of the Wahiba Sands, Sultanate of Oman. Data take 37A, 14 November 1981. The image is at L-band, HH polarisation, and has a spatial resolution of c. 40m. (Courtesy Jet Propulsion Laboratory, California)

## NEW YORK

Williams (1983) notes a similar role for Landsat imagery in a vegetated area by giving an example of northern New York state. A dark, textured, oval pattern on a Landsat image of the area defines the breached Adirondack dome, a mountainous core of fractured metamorphic rocks of Proterozoic age. This forested core contrasts with the lighter toned farmlands developed on Paleozoic strata that surround the oval uplift.

## MOHAVE DESERT

Podwysocki *et al.* (1985) used Landsat TM data of the eastern part of the Mohave desert in Nevada and Arizona to explore the value of the TM's longer wavelength bands for mineral exploration. They created a colour

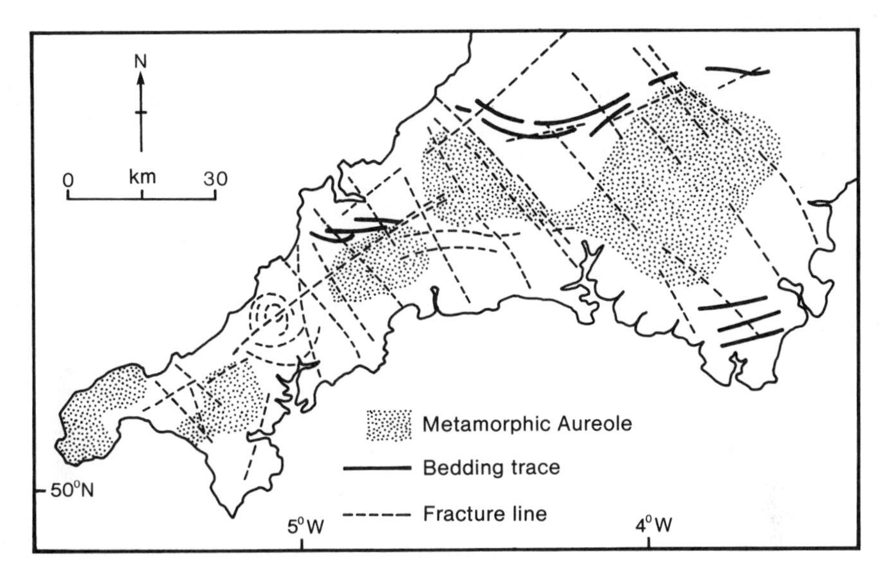


8.5 Bedding traces and fracture lines plotted from interpretation of enhanced Landsat MSS imagery, south west England. (After Moore and Camm 1982)

ratio composite (CRC) image of the area from ratio combinations of TM bands: the ratios are shown in Table 8.2. The Mohave desert study area contains Precambrian crystalline and Tertiary volcanic rocks which have been altered in part by middle and late Tertiary igneous activity which has mineralised the rocks and produced characteristic hydrothermal mineral assemblages. The limonitic materials in particular are indicators of mineralisation and most of these areas do stand out clearly on the CRC image. In the past these areas have been exploited for precious metals and are presently being re-evaluated for their precious metal and base metal potential.

## Thermal inertia

One characteristic of rocks and soils which has been useful in their analysis is their thermal inertia. This represents the ability of rocks and soil to retain their heat energy and gives insights into the composition of the surface. In April 1978 the Heat Capacity Mapping Mission (HCMM) satellite was launched to investigate the value of measuring the thermal inertia of the Earth's surface from space (Short 1982). The satellite was equipped with

*TABLE 8.2: Landsat TM ratios used by Podwysocki* et al. *(1985) for mineral exploration in the Mohave desert, USA*

| *Ratio* | *Justification* |
|---|---|
| TM5/TM7 (1.55–1.75 μm/2.08–2.35 μm) Red colour in CRC image | Distinguishes between the presence and absence of absorption bands in TM7. Hydrothermally altered rocks commonly contain concentrations of clay minerals, micas and hydrated sulphur-bearing minerals and are characterised by absorption bands in the 2.0–2.4 μm region. These rocks have a high value in this ratio. |
| TM5/TM4 (1.55–1.75 μm/0.76–0.9 μm) Green colour in CRC image | Distinguishes vegetated and non-vegetated areas. TM5 has low reflectance from vegetation because of water absorption, and TM4 has reflectance caused by the leaf's cell structure. Vegetation has a low value in this ratio. |
| TM3/TM1 (0.63–0.69 μm/0.45–0.52 μm) Blue colour in CRC image | Distinguishes materials containing limonite. Limonite typically shows a sharp reduction in reflectance from TM3 to TM1 because of a ferric iron absorption band at shorter wavelengths. Limonite yields a high value in this ratio. |

*Source*: After Podwysocki *et al.* (1985)

only one sensor, the Heat Capacity Mapping Radiometer (HCMR) which had two channels:

| | |
|---|---|
| Visible–near infrared | 0.5–1.1 μm |
| Thermal infrared | 10.5–12.5 μm |

With these two channels the HCMR could provide data on apparent albedo and on equivalent blackbody temperature.

The thermal inertia of a homogeneous material can be expressed in terms of its density, specific heat and thermal conductivity. From space there is an effect of atmospheric transmittance on the thermal inertia, so an apparent thermal inertia (ATI) has been determined (NASA 1978):

$$\mathrm{ATI} = \frac{\mathrm{NC}\,(1 - \alpha)}{\Delta \mathrm{T}} \tag{8.1}$$

where N is a scaling factor, set at 1000 to bring the values of ATI into the range 0–255 for imaging display; C is a factor related to latitude and solar declination, for example at a latitude of 45° and a solar declination of +20° then C = 1.5165; $\alpha$ is albedo, measured by the visible–near infrared channel of the HCMR; $\Delta T$ is change in temperature (deg K) from its maximum to its minimum in the diurnal temperature cycle, i.e. $T_{max} - T_{min}$.

In practice the value of $\Delta T$ cannot be measured from space as $T_{max} - T_{min}$ because the satellite passes over at a specific time which may not correspond exactly with the peak or trough of the diurnal temperature cycle. Instead, $\Delta T$ is measured as $T_{day} - T_{night}$ where $T_{day}$ is the surface temperature at the daytime pass of the satellite (c. 2 pm local time) and $T_{night}$ is the temperature at the night-time pass of the satellite (c. 2 am local time). The value of $\Delta T$ can only be calculated by geometrically correcting the day and night images to the same projection so that the differences in temperature for coincident pixels can be calculated. Even so, the pixel shapes themselves do not coincide because the directions of the spacecraft orbits for the day and night were not the same: the day orbit was inclined 11.5° west of north and the night orbit 11.5° east of north.

## GEOLOGY

Cassinis *et al.* (1984) used HCMM data of the Gulf of Orosei in eastern Sardinia to examine the relationships between rock type and ATI. They took a number of transects across their study area and compared ATI measurements from the HCMR data with geological and topographic information for transects taken from maps and field surveys. The ATI values proved to be higher on dolomites and lower on sandstones, and generally lower on eruptive rocks. There were peaks in the ATI transects at outcropping granite which may have corresponded to concentrations of silica. Cassinis *et al.* reached the following conclusions on ATI for eastern Sardinia:

1. ATI is very sensitive to lithological boundaries and to tectonic disturbances. This is caused by geological alteration and by variation in moisture content.
2. The numerical differences between the average ATI of large areas of different lithology are generally small.
3. On dry areas of uniform lithology with a thin uniform vegetation canopy, ATI anomalies caused by rock variation can be observed.

The second of these conclusions is somewhat disappointing as it suggests that rock types cannot be consistently identified by their ATI values.

However, Short (1982) has reviewed some of the HCMM investigations in the United States and reached slightly more encouraging conclusions, including:

(1) The thermal properties of rocks calculated from HCMR data can be used to distinguish hardrock units from low density rocks.
(2) Dissimilar rock types can be grouped together in classes related to their density.
(3) Some variations within the same rock type, for example basaltic texture, may be recognised.

## SOILS

While geological investigations have shown some advantages in using HCMR data, their main application has been in soil studies. In particular the role of water in soils has been studied as this has important implications both for the surface energy balance and for agricultural uses of soils. Weisnet and Matson (1983) show that there is a clear relationship between soil moisture and diurnal changes in the surface soil temperature. Figure 8.6 illustrates the linear relationship between the amplitude of the diurnal surface soil temperature wave and volumetric soil water content for the 0–2cm layer, with sample observations taken in all seasons during the period 1970 to 1973. This clear relationship suggests that thermal inertia measured from HCMR data may be useful in the measurement of soil moisture.

In Europe much of the work using HCMM data was coordinated through the TELLUS project sponsored by the European Community (Galli de Paratesi and Reiniger 1983). Fourteen test sites were used in the project and three main themes were pursued: these are briefly described below.

### *Evapotranspiration and soil moisture*

A number of models were developed to permit the calculation of cumulative daily evapotranspiration from measurements of $T_{day}$, and surface cover type from $T_{day}$ and $T_{night}$. In an operational test using HCMM data in southern France on irrigated and dry grassland areas the method agreed to within ±15 per cent with evapotranspiration measured at the surface by the Bowen ratio over a hundred-day period.

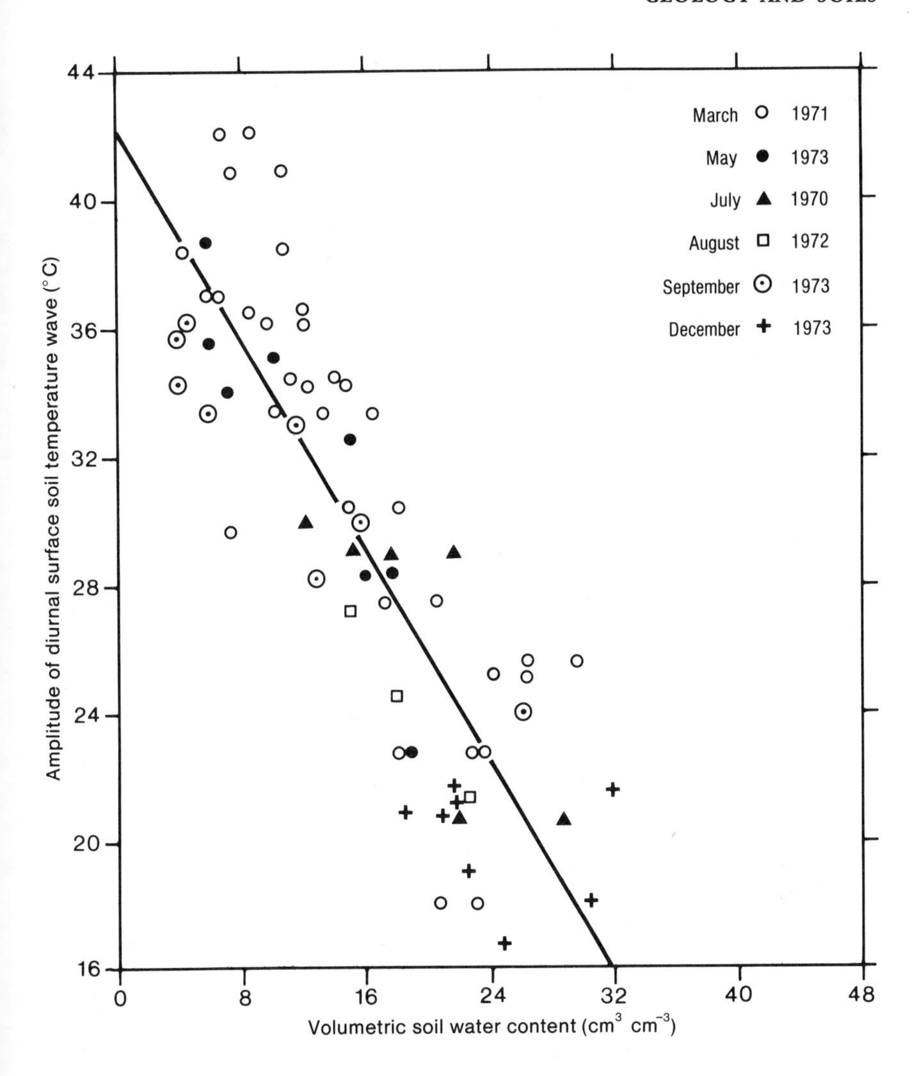


8.6 The relationship between volumetric soil water content and the amplitude of the diurnal surface soil temperature wave in the 0–2cm layer. (After Weisnet and Matson 1983)

## *Surface temperature distribution*

In this part of TELLUS the effects of topography, soils, land use and weather conditions on the surface temperature distribution were analysed. Hydrology and soil moisture were important in the determination of the surface temperature, particularly in night-time thermal infrared data. Forests and urban areas were distinguishable.

*Anthropogenic heat release*

A range of urban areas from large conurbations down to small settlements measuring 1–2 pixels in size could be identified on the HCMM imagery. The smaller settlements (<10,000 inhabitants) could only be reliably identified when the temperature contrast with their surroundings was large. The core area of every large city registered a heat island with temperatures a few degrees K higher than the surrounding: this measurement of a heat island is not the conventional one which refers to air temperature at screen level (Barry and Chorley 1982) rather than to the surface temperature.

## SURFACE CLASSES

Finally, Short (1982) has prepared a table which summarises the grey tones in HCMR images for a variety of different environments (see Table 8.3). This classification scheme is based on five forms of data: daytime visible reflectance (albedo), daytime thermal infrared ($T_{day}$), night-time thermal infrared ($T_{night}$), day–night temperature difference ($\Delta T$) and thermal

*TABLE 8.3: Classification based on grey tones in HCMR data*

| Class/feature | Day-VIS | Day-IR | Night-IR | $\Delta$T | ATI |
|---|---|---|---|---|---|
| Basalt | VD | L | M-L | MD | L-VL |
| Granite | M-L | M | L-M | MD-M | M-L |
| Soil/alluvium | M-L | L-VL | MD-M | L-VL | MD-M |
| Desert sand | L-VL | L | MD-L | ML-L | M-MD |
| Vegetation | MD-M | MD-L | M-L | MD-L | L-MD |
| Field crops | MD-D | MD-VL | MD-M | M-L | MD |
| Vegetated slopes: | | | | | |
| Sun-facing | M | L-VL | M-L | M | M |
| Shaded | MD | MD | MD | MD | L |
| Non-vegetated sheltered valleys | MD-M | M-L | L | M | M |
| Pure standing water | D | D-MD | VL | D | VL |
| Silty water | MD-L | MD-M | M-L | MD | L |
| Damp soil | MD-M | M-L | MD | M-L | MD-D |
| Urban areas | D-M | M-VL | MD-L | MD-M | L-M |
| Snow | VL | D | D | D | VL |

*Key*: VD very dark, D dark, MD medium dark, M medium, L light, VL very light; for temperature VD = cold VL = warm
*Source*: Short (1982)

inertia (ATI). The scheme in this table amounts to a five-dimensional box classifier (see chapter 6) which could be also used to classify HCMR digital data by computer. This scheme summarises the situation mainly for the eastern United States, and would require some modification for other environments.

## Soil characteristics

### PHYSICAL PROPERTIES AND SPECTRAL RESPONSE

The role of thermal inertia in soil studies is one use of satellite data in analysing soil characteristics. Most forms of satellite remote sensing data have been analysed at one time or another to examine their potential contribution to soil science. Thus far the results have been mixed, partly because the spectral responses of soils are often confused by overlying vegetation, partly because of spatial resolution limitations, and partly because the spectral response curves for soils commonly do not display strong features (Kristof and Zachary 1974, Kirschner *et al.* 1978, Thompson *et al.* 1981, Huete *et al.* 1984). However, there are some guiding principles which can be used to relate the reflectance properties of soils to their physical characteristics, at least for the visible and near infrared part of the EM spectrum (Hoffer 1978):

(1) An increase in soil moisture will cause a decrease in reflectance. This effect is most marked when the soil is dry or nearly so.
(2) An increase in particle size will cause a decrease in reflectance for a given soil type by increasing the surface roughness.
(3) An increase in organic matter will cause a decrease in reflectance: the effect is greater with greater decomposition of the organic matter.
(4) An increase in iron oxide will cause a decrease in reflectance.

These factors are often closely interrelated and not always separately represented in the field or in satellite data. Baumgardner and Stoner (1982) (cited in Milton and Webb 1984) have attempted to draw generalised curves for soils influenced by organic matter or by iron compounds. Their five curves are shown in Figure 8.7 and the label for each curve is described in Table 8.4. The wavelength range extends further than implied in the list above, and so includes water absorption features at c. 1.45μm and 1.95μm which are shown by troughs in the curves.

## SOIL TEMPERATURE AND SOIL WATER CONTENT

It was noted earlier that soil water is a very significant parameter in soil studies. Heilman and Moore (1982) examined the role of soil water in a study of the Big Sioux river basin, North Dakota, using HCMM data. The soils in the basin varied from poorly-drained silty clay loams on the flood plains to well-drained sandy loams on the terraces. Four types of data were collected:

*TABLE 8.4: Characteristic spectral response curves for the five soil types shown in Figure 8.7*

| *Label* | *Characteristic* |
|---|---|
| a | Organic dominated curve form |
| b | Organic affected curve form |
| c | Minimally altered curve form |
| d | Iron affected curve form |
| e | Iron dominated curve form |

*Source*: Milton and Webb (1984)

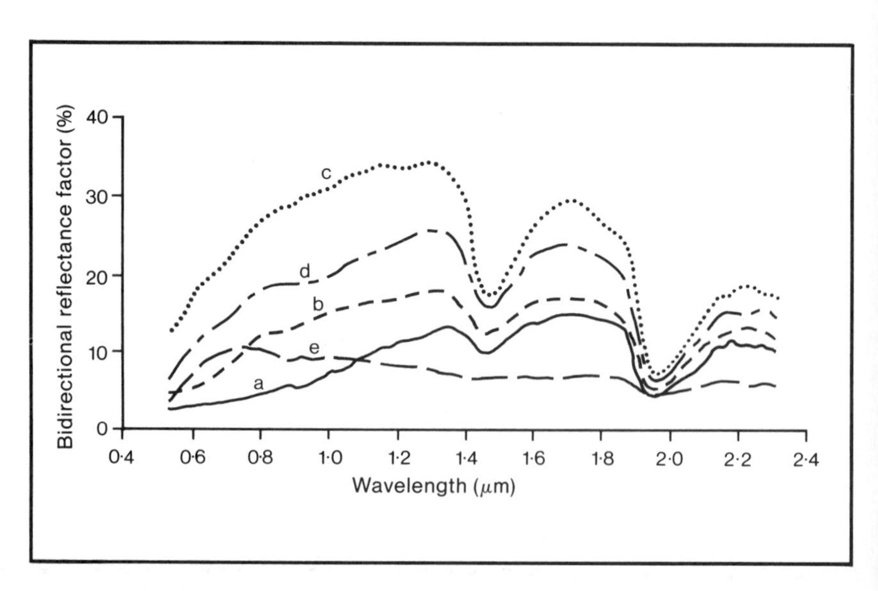


8.7 Spectral reflectance curves for five soil types. See Table 8.4 for the key. (After Milton and Webb 1984)

(1) Soil water in the 0–4cm layer measured gravimetrically from samples collected on the days of HCMM overpasses. The samples were taken from fields representative of the ranges of soils and land uses in the river basin.
(2) Percentage vegetation cover, measured by counting from a random dot grid placed over a 35mm slide projection of a surface plot.
(3) HCMR data for four overpasses in 1978 from June to September.
(4) The land use composition of each HCMR pixel determined by interpretation of a Landsat MSS image of the area taken on 13 May 1978.

The authors report that their earlier work had shown that the surface soil temperature could be related to the radiant temperature by

$$T_s = 20.35 + 0.79\ T_c e\ (-0.8P) \qquad (8.2)$$

where $T_s$ is the soil surface temperature (°C), $T_c$ is the radiant temperature of crop and soil (°C) and P is the percentage vegetation cover expressed as a fraction.

For the Big Sioux river basin, data from the thermal infrared waveband of the HCMR were used as a measure of $T_c$ and $T_s$ could be estimated from $T_c$ and P using equation 8.2. The next stage was to relate the HCMR estimate of $T_s$ to the soil water content of the 0–4cm layer (S), which gave the relationship

$$T_s = 58.35\ S^{-0.18} \qquad (8.3)$$

which had $r = 0.74$ and was significant at $\alpha = 0.01$. This study suggests that soil water content, such an important factor in soil studies, can be estimated from satellite thermal infrared data.

## MICROWAVE REMOTE SENSING OF SOILS

Microwave remote sensing from satellites is used in two forms: active and passive. Active microwave remote sensing is a radar system, normally from space a SAR, which illuminates the surface with a pulse of microwave energy which is then backscattered back to the radar antenna on the satellite (see chapter 3). Passive microwave remote sensing measures the energy emitted naturally at microwave wavelengths from surfaces. The discussion here will concentrate upon active microwave remote sensing from space, i.e. the use of SAR.

The soil parameters most frequently studied in microwave remote sensing are soil moisture, soil roughness and soil texture. An increase in soil moisture will increase the backscatter in SAR imagery. This can be seen in Figure 8.8 which shows a Seasat L-band SAR image of dry terrain in Wyoming, 1 August 1978. The bright streaks are the result of rain clouds traversing the area and wetting the ground. The ground counterpart to a SAR is a scatterometer (see Figure 4.5 in chapter 4). Batlivala and Ulaby (1977) have shown that the backscatter coefficient at C-band can be related to the soil moisture in the top 1cm of the soil, particularly when the soil moisture is expressed as a percentage of field capacity. The results of scatterometer measurements performed at the University of Kansas in 1974 and 1975 are shown in Figure 8.9. It is clear from this diagram that as soil moisture increases then the backscatter increases, precisely what is seen on the Seasat SAR image of Wyoming in Figure 8.8.

Soil surface roughness and soil texture are often related to one another. An increase in surface roughness, perhaps because of larger soil particles, can increase the backscatter on SAR imagery when the magnitude of the

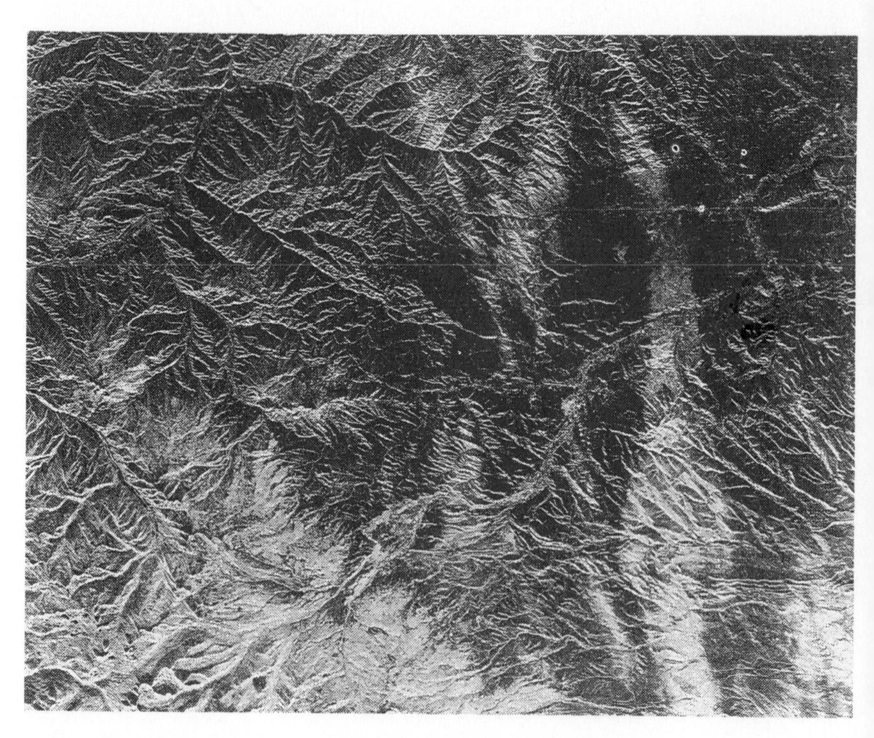

8.8 Seasat SAR image of part of Wyoming, USA, 1 August 1978. The light-toned streaks are areas of recent rainfall where the backscatter is higher as the soil moisture content has increased. (From Raney 1983)

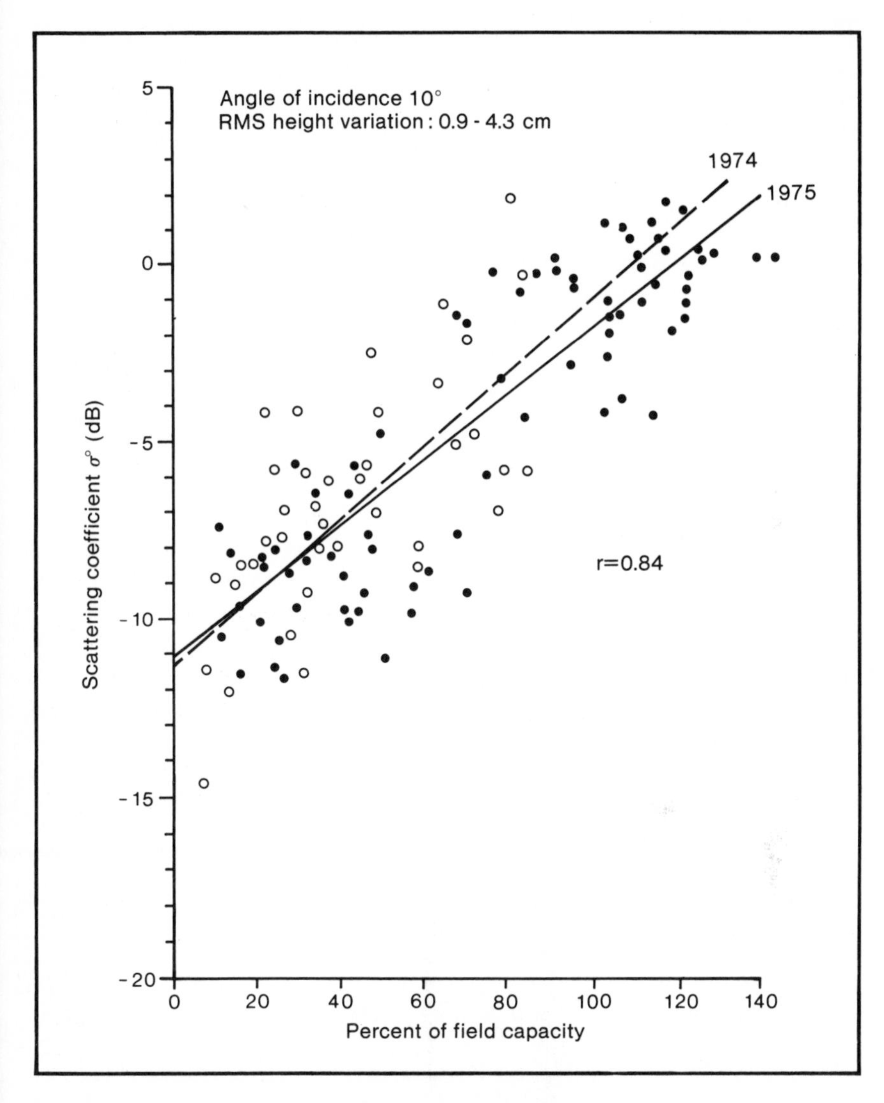


8.9 Backscatter coefficient plotted against soil moisture expressed as percentage of field capacity. The open circles are data points for 1974, the closed circles data points for 1975. (After Batlivala and Ulaby 1977)

surface roughness is greater than about half the wavelength of the radar. This means that for L-band SAR (such as Seasat or the Shuttle Imaging Radars) the horizontal magnitude of the surface roughness should be greater than c. 12cm for it to be recorded: surface roughness variation of less than 12cm will appear smooth on SAR images.

## RECONNAISSANCE SURVEYS

The visual interpretation of Landsat imagery in arid and semi-arid areas can provide useful information on soils and land surface characteristics. Mitchell (1981) gives examples of the role of Landsat MSS imagery in reconnaissance surveys of the Sudan and Jordan. For Jordan Landsat images at 1:1 million scale were interpreted and maps produced as part of a project to control soil erosion on the western plateau and to explore the enlargement of the irrigable areas elsewhere in the country. Mitchell identified land catenas and associated these with soil groups, and he stresses the importance of interpreting an assemblage of information on geology, climate and topography to assist with the interpretation of the satellite images. The approach of recognising land catenas and landscape units on Landsat imagery has also been employed in a United Nations Food and Agriculture Organisation (FAO) project to prepare a land degradation map of the world at 1:5 million scale.

Mitchell (1981) gives a second example of reconnaissance soil survey work in the Sudan savanna project, a joint project between FAO and the government of the Sudan to develop parts of the savanna zone. Again, Landsat MSS images proved useful in the identification of terrain units and provided a reconnaissance soils framework useful for separating areas of little potential value for development from those areas deserving more detailed study. The hydrographic network in the project area could also be mapped, and using wet and dry season comparisons the differences in the network on clay soils and on sands were visible.

*Chapter 9*

# Atmosphere

Having examined in the last two chapters applications of satellite remote sensing for land surface studies, this chapter turns attention to the Earth's atmosphere. The atmosphere plays an important part in affecting radiation as it travels from the sun to the Earth and then from the Earth to the satellite sensor, but in this chapter the emphasis is on the study of the atmosphere itself rather than its effects on other types of satellite remote sensing. The chapter discusses four areas of satellite remote sensing of the Earth's atmosphere: the radiation balance, rainfall estimation, airflow estimation and atmospheric sounding.

## Radiation balance

Consistent environmental information over the whole globe is notoriously difficult to obtain. Satellite data provide for the first time a consistent data set for the whole planet with which to estimate the main elements of the energy system of the globe. The net radiation of the Earth–atmosphere system measured at the top of the atmosphere is given by equation 9.1 (Weisnet and Matson 1983):

$$N = Q\,(1 - \alpha) - E \tag{9.1}$$

where N is net radiation, $Wm^{-2}$; Q is incoming solar radiation, $Wm^{-2}$; $\alpha$ = albedo, 0–1 and E is outgoing longwave terrestrial radiation loss, $Wm^{-2}$.

## SOLAR CONSTANT

The solar constant is the amount of energy received from the sun at the Earth's distance from it and at right angles to the direction of the energy flux. The Nimbus 7 satellite carried an Earth Radiation Budget (ERB) sensor (Gruber and Krueger 1984, Jacobowitz and Tighe 1984) which has allowed the estimation of the solar constant (Q) by measuring the incoming radiation to the Earth from the sun. The value of Q is $1376 Wm^{-2}$, with an RMS deviation of $\pm 0.73 Wm^{-2}$: thus the solar constant is not constant but varies over time.

## ALBEDO

The albedo ($\alpha$) is the proportion of the incoming radiation reflected back to space and is expressed in the range 0–1, where 1 is total reflection. Albedo is the full spectral reflection in the wavelength range 0.2–4.0$\mu$m, and has been estimated using data from (1) the 0.2–3.8$\mu$m channel of the Nimbus 6 satellite with data collected from twenty-four months during the period 1975–7, and (2) the 0.5–0.7$\mu$m channel of Noaa satellites 2–5 for forty-five months for the period 1974–8 (Henderson-Sellers 1984). The Noaa data provide only a narrow band of the full spectral reflection, but this part of the EM spectrum (0.5–0.7$\mu$m) corresponds with the maximum solar energy input and provides a good estimate of albedo (see Figure 2.4 in chapter 2). The annual and global averaged albedo is currently estimated as 0.306.

## TERRESTRIAL RADIATION

The outgoing long wavelength terrestrial radiation loss (E) can be estimated by using data from thermal infrared wavebands. Regression equations have been employed (Weisnet and Matson 1983) to estimate the total outgoing terrestrial radiation by using the 10.5–12.5$\mu$m channel data from the Noaa 2–5 satellites. These regression equations considered ninety-nine different model atmospheres with a range of temperature, moisture and cloud conditions: the regression models explained 98 per cent of the variance of E from the Noaa thermal infrared data.

## NET RADIATION

The variables on the right-hand side of equation 9.1, that is Q, α and E, can be measured or estimated using satellite data and so the net radiation at the top of the atmosphere (N) can be estimated by using this equation. However, with polar orbiting satellites only a relatively small data sample can be obtained because of the orbit characteristics. Noaa satellites pass over the same point on the Earth's surface twice a day (one in daylight and one at night) so measurements of E can be taken twice per day, but measurements of α can only be taken during daylight. This clearly restricts the time at which data can be collected and so the representativeness of the calculations of net radiation.

Figure 9.1 shows the annual cycles of albedo, outgoing terrestrial radiation and net radiation calculated from the forty-five month (1974–8) Noaa data set referred to earlier (Henderson-Sellers 1984). The albedo exhibits a pronounced annual cycle, with the lowest values in the northern hemisphere summer when the vegetation growth is at its most extensive and cloud cover is low. The outgoing terrestrial radiation curve shows a reverse pattern, with a maximum during the northern hemisphere summer. The net radiation curve has the same phase as the albedo curve and reaches a

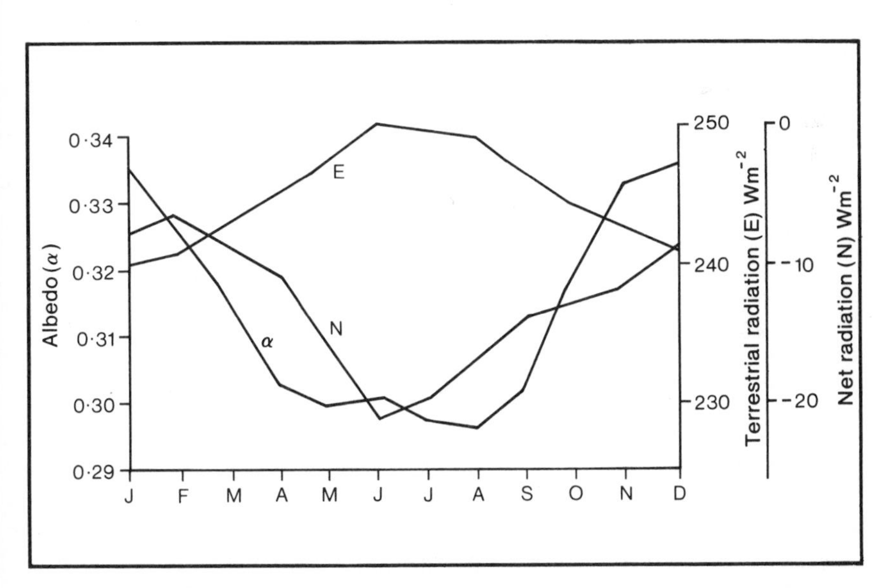


9.1 Annual cycles of albedo (α), outgoing terrestrial radiation (E) and net radiation (N) of the Earth–atmosphere system. Based on forty-five months of Noaa data, 1974–8. (After Henderson-Sellers 1984)

minimum in the northern hemisphere summer when the Earth is furthest from the sun.

The spatial variation in net radiation is shown in Figure 9.2. This map, for the zone 60° N to 60° S, shows positive net radiation for most of the tropical zone and the southern hemisphere continents, with the important exceptions of the Sahara and Arabian deserts. These deserts are sinks of radiation because they have a high albedo and so reflect a large proportion of the incoming solar radiation, and as they have low cloud cover they are areas of high outgoing terrestrial radiation loss. Almost everywhere poleward of 40° latitude has a net radiation loss, reaching a maximum over Greenland (not shown) of below $-150\mathrm{Wm}^{-2}$. The net radiation distribution shown here explains why the general circulation of the atmosphere attempts to redistribute energy polewards, to transfer energy from the tropical areas of surplus to the remaining areas of deficit.

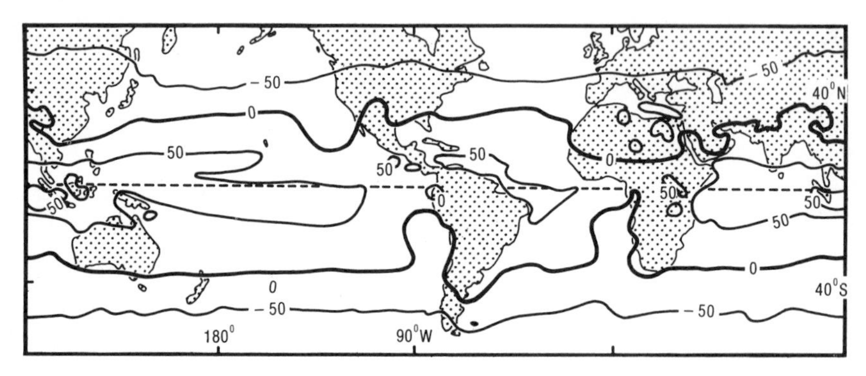


9.2 The net radiation pattern for the Earth between 60° N and 60° S ($\mathrm{Wm}^{-2}$). The data are from the same source as Figure 9.1. (Redrawn from Weisnet and Matson 1983)

## Rainfall estimation

One advantage of satellite remote sensing is that it provides data for remote areas. In the case of rainfall data there are many data voids or data sparse areas on the globe, particularly the oceans, the deserts, some continental interiors, and mountain regions. Clearly satellite remote sensing has an attraction if it can assist in providing information on rainfall for such areas as well as aiding rainfall estimates in other areas. In pursuing rainfall estimation from satellites a wide variety of satellite data have been used, including visible and near infrared, thermal infrared and microwave wavelengths. Atlas and Thiele (1981) and Barrett and Martin (1981) give reviews of the range of methods and wavelengths used.

## CLOUD BRIGHTNESS

One early proposition in this field was that the brightest clouds seen on visible wavelength satellite images are those most likely to be precipitating (Woodley and Sancho 1971). This is particularly the case with cumulonimbus clouds in the tropics. So, if a contour map of cloud brightness is produced then it may be possible to convert it to a map of raining clouds. Work of this type was undertaken at the NOAA Experimental Meteorological Laboratory in Florida by analysing data from the ATS 3 geostationary satellite in conjunction with data from a ground radar calibrated with a raingauge. The radar used was a plan position indicator (PPI) radar which provides a strong radar echo from those clouds which are raining (Wilson and Roesli 1985). Woodley and Sancho (1971) found that in their measurement period 322 of the 372 radar echoes analysed fell within areas of high cloud brightness on the ATS 3 data, and that 87 of the 124 ATS 3 brightness centres encompassed radar echoes. Some of the discrepancies were caused by the age of the cloud: young bright clouds were found to be more likely to be precipitating than old bright clouds. The methods have been developed further and applied to problems of rainfall estimation in Venezuela, West Africa and the central Atlantic ocean (Griffith *et al.* 1978, Woodley *et al.* 1980).

Another approach using cloud brightness has been developed by Oliver and Scofield (1976), although their method has not received wide attention. In their method cumulative brightness is calculated for fixed ground positions by summing the image brightness for a series of thermal infrared geostationary satellite images of an area. Typically brightness is summed for each one degree latitude/longitude intersection in a study area for the twelve geostationary images in a six-hour period. The cumulative brightness values are then correlated with surface raingauge data so that the map of cumulative brightness can be converted into a map of rainfall. In a study of part of the Pacific Ocean Oliver and Scofield found the following relationship

$$y = 0.0185x \qquad (9.2)$$

where y is the observed six-hour precipitation and x is the cumulative six-hour thermal infrared image brightness. This relationship of brightness with rainfall had a correlation (r) of 0.63. When only data for the upwind, active position of the cloud systems were analysed the correlation increased to 0.88.

## AREAL STATISTICS

For more than a decade and a half Eric Barrett has developed and modified a method for estimating rainfall based on cloud area and cloud type information from weather satellite data. The basic form of his areal statistics method is

$$R = f\,(c, i, a) \tag{9.3}$$

where R is the accumulated rainfall for a ground point or ground area, c is the cloud area, i is the cloud type and a is the altitude of the surface. This approach is based on the two premises that (1) as cloud amount increases then the probability of rainfall increases, and (2) some cloud types have a greater propensity to produce rainfall than others.

In the Barrett method the study area is divided into a grid of cells, some of which include ground raingauges and many which do not. For a certain time period (say a week) the data on cloud area/amount and cloud type are summed for each cell and a regression relationship produced between the cloud data and the ground rainfall data, as in equation 9.3. The regression model, calibrated in this way with the ground data, can be applied to those grid cells without raingauges to estimate the rainfall for each cell from the satellite data alone.

This method of estimating rainfall has been used by Barrett in studies of the tropical Far East, Oman and North Africa (Barrett 1970, 1971, Barrett and Martin 1981). The method has been modified over time to take into account the characteristics of the areas for which it has been used and the needs of those using the results.

A similar method to that developed by Barrett has been used for meteorological rather than climatological purposes. Follansbee (1973) estimated rainfall for a twenty-four hour period in Zambia, and later the method was applied in the USA, the USSR and China. In his model Follansbee estimated total twenty-four hour rainfall (R) as

$$R = (b_1x_1 + b_2x_2 + b_3x_3)\,/b_0 \tag{9.4}$$

where $x_1$, $x_2$ and $x_3$ are areas of the total study area $b_0$ covered by the most important types of rain-producing clouds, viz. cumulonimbus, cumulus congestus and nimbostratus. The values of $b_1$, $b_2$ and $b_3$ are calculated by regression methods.

A number of other methods based on areal statistics have been proposed and Atlas and Thiele (1981) summarise their form as a multiple regression model:

$$R = b_0 + w\sum_{i=1}^{n} b_i f_i \tag{9.5}$$

where R is rainfall; w is a weighting derived from other sources, for example altitude; $f_i$ represents the n independent variables such as cloud brightness, area or temperature and $b_0$ and $b_i$ are constants fitted by calibration of the model.

Although most of the areal statistics models have used a multiple regression approach, either explicitly or implicitly, few studies have fully considered the constraints placed on the analysis by the statistical assumptions which lie behind this class of models.

## TEMPORAL SEQUENCES

One way of utilising the frequent cloud images produced by the geostationary weather satellites is to monitor cloud growth and relate this to precipitation. The method developed by Scofield and Oliver (1977) offers weather forecasters a way of measuring convective rainfall by using enhanced thermal infrared geostationary satellite imagery. The method differs from those discussed above by being less quantitative and more based on interpretation. For a target ground location on an image three questions are answered by the forecaster:

(1) Is the cloud convective?
(2) Is the cloud cold?
(3) Is the target ground location under the active part of the cloud, as identified by towering cloud cells and low temperatures?

If the answers to all three questions are positive then significant rainfall is predicted for the target ground location. There are then two further questions which help estimate the amount of rainfall.

(4) Is the cloud growing?
(5) Are there overshooting tops or merging convective towers?

From this description of the significant cloud characteristics in a sequence of geostationary satellite images, rainfall rates are assigned. The possible rain rates fall into forty-eight categories which range from trace rainfall to 178mm per hour. The method has proved particularly useful for estimating storm rainfall, and has been applied in Iowa, North and South Carolina, Venezuela and Colombia.

## MICROWAVE METHODS

The three groups of methods discussed above use visible and/or thermal infrared imagery as their basis. The principal problem with this is that there is no clear physical mechanism which relates cloud brightness and cloud top temperature to the physical processes of rainfall, although clearly there are some statistical relationships. This problem can be answered by looking at the microwave part of the EM spectrum. At microwave frequencies the absorption and scattering of EM energy both increase as the rainfall rate increases because of the physical properties of the rain droplets. Wilheit *et al.* (1977) showed a reasonable relationship between rainfall rate and the brightness temperature measured at 19.5GHz by the Electrically Scanning Microwave Radiometer (ESMR) on Nimbus 5. Figure 9.3 shows this relationship, which was calibrated using a meteorological radar located in Miami. The relationship is log-linear for higher rainfall rates, but for low rainfall rates (<3mm per hour) the microwave brightness temperature is not sensitive to changing rainfall rates.

The relationship of microwave brightness temperature and rainfall rate works best over ocean areas because of the contrast between the low brightness temperature values of the raining areas of clouds and the higher values of the sea surface and non-raining cloud areas. Over land, rain patterns can be discerned on ESMR data, but there is confusion with areas of high soil moisture content. There is some promise that higher frequencies (i.e. shorter wavelengths) may provide better discrimination over land, and at 37GHz such as the ESMR on Nimbus 6 it is possible to classify the data into rain, dry ground and wet ground.

## EXTENSION OF GROUND RADAR

In this section on rainfall estimation a number of references have been made to the use of ground-based PPI radars which are calibrated to identify raining clouds. The UK Meteorological Office was the first national organisation to develop an operational system which combines ground-based radar with satellite data, so that the inferences derived from the radar information could be extended to cover larger areas. Most of England and Wales is covered by four ground-based radars (see Figure 9.4) whose data are combined and remapped onto a conventional map projection. The radars are calibrated with raingauge data and have been very useful in tracking the development of mesoscale rain systems in real time over the UK (Browning 1985). The spatial coverage of the data is extended by embedding the radar maps in Meteosat visible and thermal infrared image data and extrapolating the rain areas from the radar coverage area to the Meteosat coverage area of north-west Europe (Browning and Collier 1982). Forecasts of rain movement (Schlatter 1985) can be generated by an extrapolation of the radar–satellite data because of the high temporal frequency of the data.

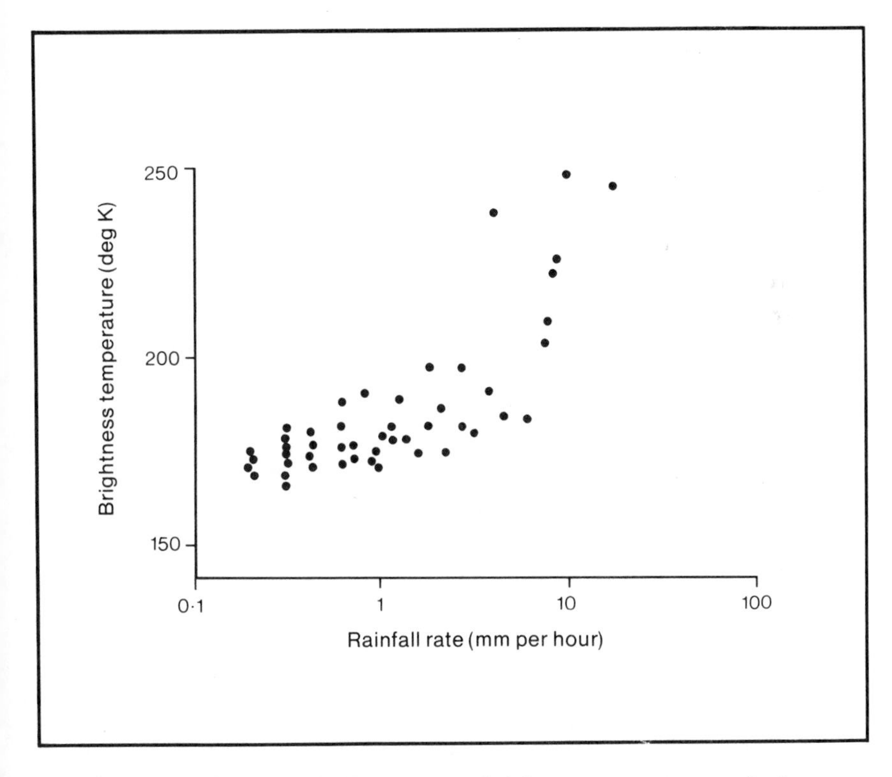


9.3 Relationship between ESMR microwave brightness temperature and rain rate. (After Atlas and Thiele 1981)

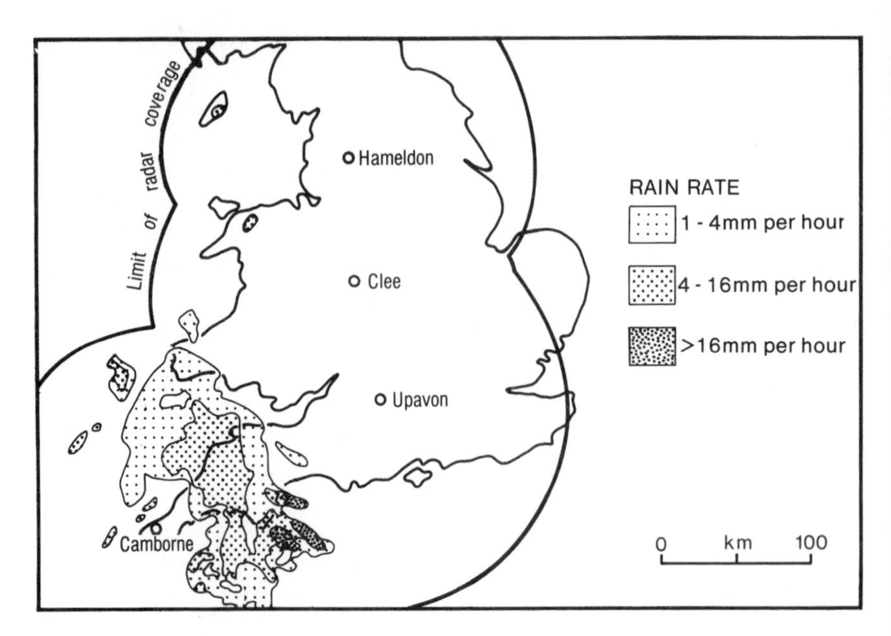


9.4 Location of four UK ground-based radars and the boundary of their coverage. A rain area over south-west England is shown. (Redrawn from Browning 1985 and Browning and Collier 1982)

## Airflow estimation

The high temporal frequency of geostationary satellite images is also useful in following the movement of clouds which are in turn related to wind direction and wind speed (Smith and Kelly 1985). A sequence of geostationary images has a number of benefits for airflow estimation:

(1) Determine wind direction and speed near the surface and in the upper atmosphere for areas with poor conventional meteorological data such as the oceans.
(2) Identify areas of surface convergence and divergence and thereby the areas of potential instability and thunderstorm development.
(3) Forecast the movement of weather systems such as tropical hurricanes and storms.
(4) Provide information for the production of pressure pattern charts which are at the heart of much meteorological description, explanation and forecasting.

## BRIGHTNESS CENTRES

The primary approach used in airflow estimation is to track cumulus clouds identified as brightness centres on geostationary satellite images. If the same cloud can be identified on successive images then a motion vector can be mapped which shows the direction and the distance the cloud has moved in the time interval between the two images. If many clouds are tracked then this can give a good indication of regional airflow patterns. Small cumulus clouds with tops below 3km give a good indication of the airflows in the lower troposphere, while cumulus congestus clouds with tops at 8–10km give information on the airflow in the upper troposphere. The assumption here is that cumulus clouds move at the same speed as the air, which is a reasonable assumption for most cumulus clouds, although on occasions the cloud speed can be slower than the wind speed.

## IMAGE CORRELATION

A second technique of cloud tracking is termed image correlation, and is summarised in Figure 9.5. Here a cloud area is identified within a box ($b_1$) on an image taken at time $t_1$ (Figure 9.5(a) ). The image at time $t_2$ (say thirty

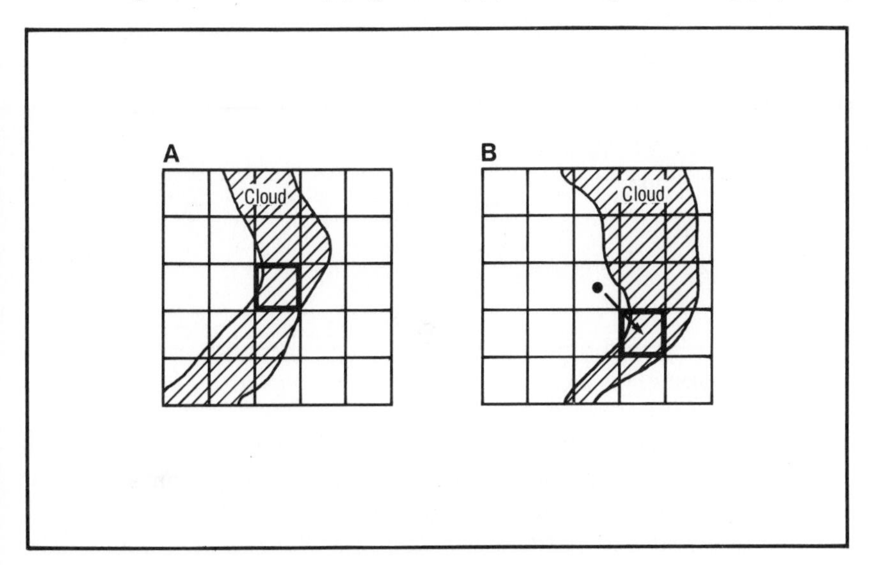


9.5 Image correlation. (a) Box $b_1$ marked bold on image at time $t_1$. (b) Box $b_2$ marked bold on image at time $t_2$. Box $b_1$ has the highest correlation with box $b_2$ on image at $t_2$

minutes later) is searched to find the box which has the highest correlation with the target box on the first image: this is shown as box $b_2$ on Figure 9.5(b). When the $b_2$ box is found then the cloud movement is identified as from the centre of box $b_1$ to the centre of box $b_2$. As with the brightness centre approach, this method can be repeated for a number of boxes thereby generating a number of cloud motion vectors. The method can work well if cloud areas retain their shape from one image to the next, but if the shape changes significantly then correlations are poor and incorrect movement may be identified.

## APPLICATIONS

Peslen (1980) adopted both the brightness centre and the image correlation methods in a study of low-level (1.0–2.8km) cumulus clouds over the central USA. Peslen used five-minute interval SMS 2 data and compared her results for wind speed with measurements from rawinsonde vertical ascent data from Topeka, Kansas and Monette, Missouri. The data were then used to estimate areas of divergence and convergence, and through the convergence estimates to identify zones of severe thunderstorm development. The differences in wind velocity between the rawinsonde data and the satellite derived data are shown in Table 9.1 for 950mb and 800mb heights. The comparison for Monette is reasonable at both heights, but a poor performance is found at Topeka.

*TABLE 9.1: Differences in velocities of winds derived from SMS 2 satellite data and those measured by rawindsonde ascents*

| *Station* | *$\Delta V$ 950mb ($ms^{-1}$)* | *$\Delta V$ 800mb ($ms^{-1}$)* |
|---|---|---|
| Topeka, Kansas | 3.7 | 4.1 |
| Monette, Missouri | 1.1 | 1.0 |

*Source*: Peslen (1980)

Johnson and Suchman (1980) used the brightness centre method for wind estimation because of problems with cloud shape changes. For a study area in the central USA of 6° latitude by 8° longitude they were able to improve on the quantity of wind estimates from just three using rawinsonde data alone to (1) seventeen when thirty-minute interval GOES data were used and (2) 165 when three-minute interval GOES data were used. Johnson and Suchman concluded that thirty-minute interval data are satisfactory for tracking clouds at the cirrus level at the top of the troposphere when clouds change only slowly, but for low level clouds images at six- to ten-minute intervals are desirable.

## WATER VAPOUR BUT NO CLOUDS

Tracking cloud movement is clearly only possible where clouds are present, but what happens in areas without clouds such as the large sub-tropical high pressure areas of the globe where cloud development is normally limited? Kästner *et al.* (1980) used an alternative to the normal visible and thermal infrared data by analysing data from the Nimbus 5 6.3μm water vapour channel. Although they took polar orbiter satellite data the approach was taken partly as preparation for the water vapour channel on Meteosat. The advantage of this channel for cloud tracking is that while clouds may be absent from an area, water vapour concentrations are high enough over most of the globe to provide information on the spatial variation of water vapour patterns.

The Kastner *et al.* approach was one not of cloud tracking but of contour tracking. They took successive Nimbus 5 6.3μm channel images and on each identified the sharp boundaries between areas of different water vapour concentration. The boundaries on two successive images could then be compared and the relative displacement calculated. The results from this work showed errors of 5–6$ms^{-1}$ compared with radiosonde data. This is unacceptable for operational meteorology, but the images used were twelve hours apart and the prospects for improvement were thought to be great.

## OPERATIONAL MEASUREMENTS

By 1982 GOES and Meteosat satellites were providing visible, thermal infrared and water vapour data operationally. In Europe Meteosat data are used operationally at the European Space Operations Centre (ESOC) in Darmstadt to provide airflow estimates. The satellite cloud winds measured at ESOC are on average 1.6$ms^{-1}$ below wind speeds measured by other observation sources, partly caused by the problems of semi-transparent cirrus clouds (Schmetz and Turpeinen 1986). Smith and Kelly (1985) show that in the USA cloud tracking and water vapour tracking are combined operationally to estimate winds: they give an example for 26 April 1982 where radiosonde data provided twenty-three measurements of wind direction and speed over much of the USA, and GOES data provided over two hundred measurements.

## Atmospheric sounding

Most of this book is concerned with relatively high spatial resolution satellite remote sensing of surfaces, be they land, water, ice or cloud. One operational branch of satellite remote sensing is not concerned with measuring such surfaces, but with measuring the characteristics of the atmosphere vertically (Houghton *et al.* 1984). The spatial resolution of such systems is low, typically tens of kilometres, but as the conventional network of atmospheric sounding is relatively limited then this spatial resolution is a significant step forward in collecting information about the vertical structure of temperature or of other variables in the atmosphere (Houghton 1985).

The Tiros-N satellites (see chapter 5) carry a High Resolution Infrared Sounder (HIRS2) and a Microwave Sounding Unit (MSU) (Jerrett *et al.* 1982, Susskind 1984). The HIRS2 instrument has nineteen channels and the MSU four channels. Seven of the HIRS2 channels are in the 15μm carbon dioxide absorption band and five in the 4.3μm carbon dioxide band. Each channel has a weighting function which makes it sensitive to a layer of the atmosphere. Figure 9.6 shows the weighting functions for five of the HIRS2 channels and two of the MSU channels. Each weighting function peaks at a certain layer of the atmosphere, and by using regression techniques the radiance for each channel can be related to the temperature of that layer of the atmosphere. For example, on Figure 9.6 the M4 curve (i.e. MSU channel 4) peaks at around the 70mb layer of the atmosphere, while the H15 curve (HIRS2 channel 15) peaks at around 675mb. The MSU channel 4 data can therefore be used to provide data on the temperature at the 70mb level, and the HIRS2 channel 15 data to give temperature data for the 675mb level.

Different groups of channels of vertical sounder data allow the determination of different features of the atmospheric temperature profile. The 15μm channels allow the calculation of the broad temperature structure from the surface to 20mb (the lower stratosphere), while the 4.3μm channels have peaks that are concentrated in the lower troposphere and so provide greater sensitivity in atmospheric temperature measurement close to the ground. The 15μm and 4.3μm channels are all affected by the presence of cloud, so in cloudy conditions the MSU microwave channels are used to provide information on vertical temperature structure. The four MSU channels have frequencies between 50.3 and 57.95 GHz (c. 0.5cm wavelength), and although they are not affected by cloud they have a lower resolving power in the vertical and horizontal compared to the HIRS2 infrared sounder. The sub-satellite resolution of the MSU is 100 × 170km, while the HIRS2 is 20 × 42km.

The data from the HIRS2/MSU instrument have been used by the UK

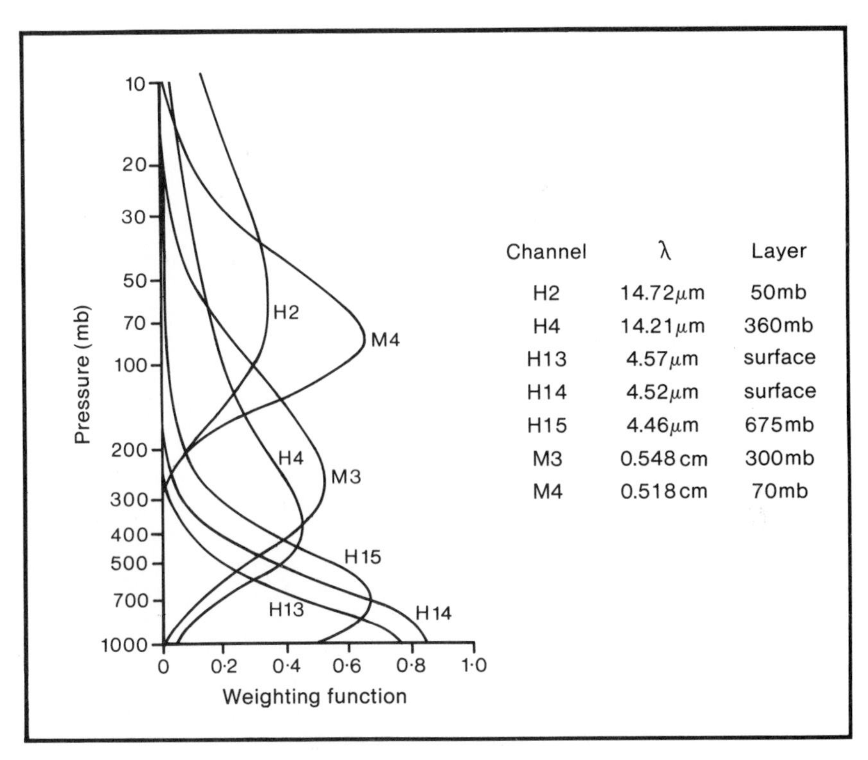


9.6 Weighting functions for five HIRS2 channels (H) and two MSU channels (M) carried on Tiros-N satellites. The wavelength (λ) and the layer of the atmosphere whose temperature is measured are also given. (After Susskind 1984)

Meteorological Office to provide temperature data on the 1000–500mb lower layer of the atmosphere. The thickness of this layer is proportional to its temperature and the thickness pattern can be used by the forecaster to predict the speed and the development of weather features (Jerrett *et al.* 1982). The vertical sounder data are particularly useful over the North Atlantic where radiosondes are sparse, and the data allow the refinement of thickness charts or the redrawing of the charts in zones with no conventional data. The HIRS2/MSU data are also used by the Goddard Laboratory for Atmospheric Sciences (GLAS) to compute atmospheric temperature profiles, cloud fields, surface emissivity and surface temperature (Susskind 1984).

## Weather forecasting: a note

The value of satellite remote sensing data for operational meteorology has been questioned in certain quarters. Houghton (1985) reports on work at the European Centre for Medium Range Weather Forecasting (ECMWF) which explored the contributions by different forms of data to the predictive skill of numerical weather forecasts.

The ECMWF took data for the First GARP Global Experiment (FGGE) and examined the weather predictability in days in three categories: good forecasts, useful forecasts and possibly useful forecasts (see Bengtsson (1982) for details). Having established these figures for all the FGGE data, the ECMWF then extracted different types of data from the FGGE data base and recalculated the weather predictability. Some of the results are shown in Table 9.2: part (a) is for the northern hemisphere and part (b) for the southern hemisphere. Using all FGGE data (System A) it was possible to produce good forecasts for 4.4 days ahead for the northern hemisphere and 3 days ahead for the southern hemisphere. If all aircraft and satellite data

*TABLE 9.2: Weather predictability for (a) the northern hemisphere and (b) the southern hemisphere in three categories of forecast. Figure are in days*

| *System* | *Data base* | *Good forecasts* | *Useful forecasts* | *Possibly useful forecasts* |
|---|---|---|---|---|
| *(a) Northern hemisphere* | | | | |
| System A | All FGGE data | 4.4 | 7.0 | >10 |
| System B | FGGE data less aircraft data | 4.4 | 6.9 | >10 |
| System C | FGGE data less satellite data | 4.2 | 7.0 | 9.8 |
| System F | FGGE data less aircraft and satellite data | 3.4 | 5.5 | 7.5 |
| *(b) Southern hemisphere* | | | | |
| System A | All FGGE data | 3.0 | 5.5 | 7.2 |
| System B | FGGE data less aircraft data | 2.9 | 5.5 | 7.2 |
| System C | FGGE data less satellite data | 1.6 | 3.5 | 5.6 |
| System F | FGGE data less aircraft and satellite data | 1.6 | 3.5 | 5.6 |

*Source*: Houghton (1985)

are removed from the database (System F) then the predictability drops to 3.4 and 1.6 days respectively. In both hemispheres the effect is significant and, as expected, is greater in the southern hemisphere where aircraft data play a less important role.

# *Chapter 10*

# Hydrosphere

In this chapter we return to the surface of the planet and examine the water bodies, the hydrosphere, of the Earth. That is not to neglect the important links between the atmosphere and the hydrosphere, and ESA (1985) neatly summarises the link between the two and why it is important to study the hydrosphere:

> More than 70% of the Earth's surface is covered by water in the form of water or ice. An understanding of the ocean- and ice-covered regions of the globe and their interactions with other components of our planet is therefore essential for the study of climate for the exploitation of resources in and beneath the oceans.

This chapter addresses the hydrosphere by discussing sea surface temperature, the physical structure of the ocean and the coastal seas, water quality, fishing applications, surface biology and snow. These form important elements of satellite remote sensing of the oceans and of the ice and snow areas (Hall and Martinec 1985, Maul 1985).

## Sea surface temperature

One parameter of the oceans which has both scientific and commercial importance is the sea surface temperature (SST). Two principal wavebands have been employed to measure SST from satellites: thermal infrared and microwave.

## THERMAL INFRARED

Clearly the thermal infrared channels of satellite sensors are appropriate instruments for measuring SST, and Callison and Cracknell (1984) give examples of water surface temperature measurement around Britain. Harries *et al.* (1983) used data from the 3.7μm, 11μm and 12μm channels of the Noaa 7 AVHRR (see Table 5.4 in chapter 5) to estimate SST, and compared their measurements with *in situ* ship and buoy measurements of SST. The comparison is shown in Figure 10.1. The comparisons with surface data were primarily made for the region 37° N to 65° N, with three additional data points (>25 °C) in the tropics. The standard deviation of the difference between the satellite-derived SST and the SST measurements *in situ* was 0.88K.

NOAA issues maps of SST operationally under the name MCSST (Multi-Channel Sea Surface Temperature) (Robinson *et al.* 1984). This product superseded the maps of GOSSTCOMP (Global Operational Sea Surface

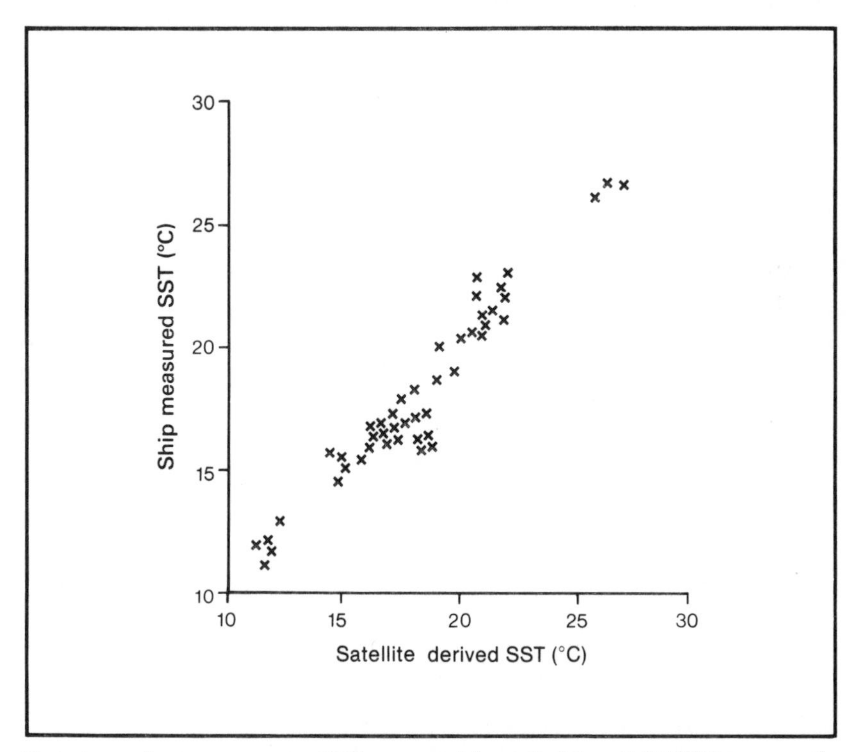


10.1 Sea surface temperature (SST) measured from the Noaa 7 AVHRR compared with ship and buoy data. (Redrawn from Harries *et al.* 1983)

Temperature Composite) data which used only one AVHRR thermal infrared channel. The use of two or three AVHRR channels for the MCSST maps together with information on atmospheric composition for correction purposes has resulted in improved data with a standard deviation of the differences between satellite and *in situ* data of the order of 0.6K. This is still too poor for some applications which require accuracies of less than 0.2K, but the value of 0.6K may result from the different methods of measurement employed *in situ* and by satellite. The ship or buoy measurement of SST is performed by a bulk measurement of the top layer (c. 1m) of the water. The satellite measurement of SST on the other hand is a measure of the water's skin temperature in a layer of negligible depth. This incompatibility may explain the mismatch between the two estimates of SST and asks the question: what is meant by sea surface temperature?

In a review of the subject, Robinson *et al.* (1984) suggest that the bulk SST of a layer of water below the skin is the most useful estimate of the true sea surface temperature, particularly for climate modelling and air–sea interaction studies, and so the satellite-derived estimates of SST should be corrected to account for the different measurement method.

The image shown in figure 10.2 is a Noaa 7 thermal infrared image (10.5μm–11.5μm) of the western North Atlantic (see also Needham 1984). This image is one of thermal radiance data, but by using a calibration model such as shown in Figure 10.1 the data could be converted to SSTs. The image shows the warmer waters of the Gulf Stream (darker tones) meeting the colder northern waters (lighter tones) off the eastern seaboard of the United States. Imagery such as shown in Figure 10.2 is used daily by NOAA oceanographers to plot the course of the Gulf Stream and its eddies: this information is provided daily to the fishing industry and to shipping concerns. Zheng *et al.* (1984) used such thermal infrared imagery for the period 1977–82 to study the dynamics of the Gulf Stream eddies. They were able to calculate the area of the rings and their rotation speeds, and estimate the horizontal eddy viscosity of the rings. The average diameter of the eighteen rings studied was 172km and the average rotation period was fourteen days.

## MICROWAVES

Data from the Scanning Multichannel Microwave Radiometer (SMMR), carried on both the Nimbus 7 and the Seasat satellites, have been analysed to estimate SST. Harries *et al.* (1983) report a study by Bernstein which compares SST maps generated from Seasat SMMR data with ship data for the north-west Pacific Ocean. The maps produced from the two data sets agree in the main structures, and the standard deviation of the difference of the two estimates was 0.75K. The SMMR estimates showed a consistent warm bias of 0.22K.

## Physical structure

The measurement of the sea surface temperature from satellites can also be used to provide information on the physical structure of the ocean and coastal seas. Zheng and Klemas (1982) used winter season thermal infrared imagery of the Yellow Sea and East China Sea from a number of different satellites, including Noaa, DMSP and GMS 1 (see chapter 5). The thermal infrared data were calibrated with *in situ* ship measurements to produce SST maps of the study area. A series of these maps was used to analyse the

10.2 Noaa 7 AVHRR thermal infrared image (10.5–11.5μm) of the Gulf Stream in the North Atlantic, 25 April 1982. (Courtesy NOAA)

location of temperature fronts and their related water currents: a summary map of these features is shown in Figure 10.3. Warm water currents from the south meet cold water currents from the north at a central ring eddy and at three main fronts running parallel to their respective coasts: Shandong, Korean and Zhejiang–Fujian. The structure revealed on this map was deduced from the temperature gradients and spatial patterns evident on the satellite thermal infrared imagery and SST maps.

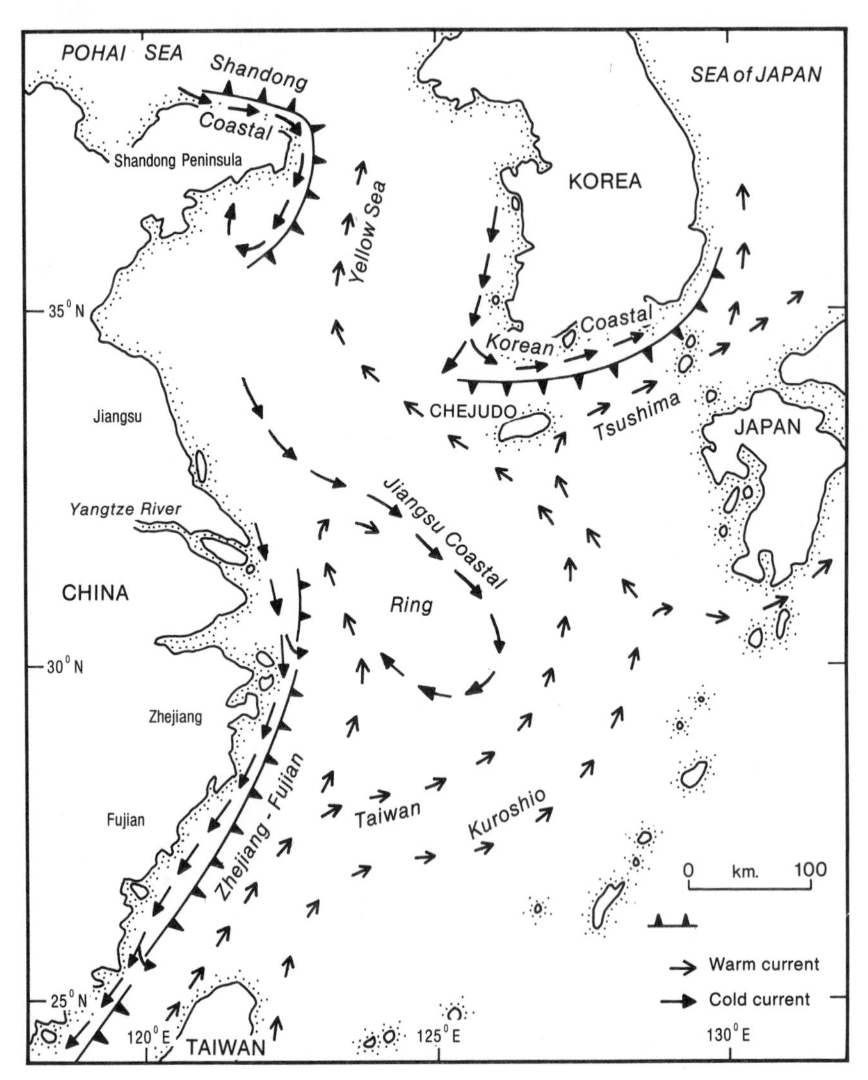


10.3 Winter surface currents in the Yellow Sea and East China Sea determined from sea surface temperature patterns. (Redrawn from Zheng and Klemas 1982)

Klemas (1980) used a series of thirty-six Landsat MSS images to chart the location of several types of fronts during twelve one-hour segments of the tidal cycle in Delaware Bay on the eastern seaboard of the USA. The location of the fronts was related to the water turbidity patterns which were estimated primarily using Landsat MSS band 5. This information was then combined with National Ocean Survey tidal current charts to help interpret the water current flow related to the fronts. One objective of the research reported by Klemas was to study the movement of oil slicks (Asanuma *et al.* 1986) and the dispersion of other pollutants (see also Massin 1984), and as this movement is closely related to frontal and current movement then the detailed knowledge provided by Landsat images can be used to assist the operational control of water pollution in Delaware Bay. Ackleson *et al.* (1985) also used Landsat MSS imagery of Delaware Bay and compared these data with SPOT simulation data.

Another research programme located off the eastern seaboard of the USA is reported by Mattie *et al.* (1980). They concentrated on the use of Seasat SAR imagery for a region off Cape Hatteras in the western North Atlantic. The Seasat images were processed to give information on the direction and length of the principal ocean wave trains, which had significant wave heights as low as 1m and wave periods of fourteen seconds. In addition, wave diffraction, bottom sand waves and internal waves were evident on the images. In the case of these SAR images the image response was controlled by backscatter, which can be related to variations in ocean wave and water microtopography, salinity and the presence of debris on the water surface. The observation of sub-surface features is possible using Seasat SAR images by a knowledge of the physical relationships between the sub-surface wave structures and surface wave patterns. Lodge (1983) gives examples of wave refraction patterns and deep water features around Britain visible on Seasat SAR images.

The relatively short flight of Seasat in 1978 has provided a great deal of information on the physical characteristics of the oceans. The work of the Seasat Users Research Group in Europe (SURGE) is brought together in a collection of papers edited by Allan (1983). As Seasat was equipped with a scatterometer, an altimeter and a SAR, the data have been processed in a number of ways to map wind speed and direction at the ocean surface, significant wave height, eddy energy, the topography of the ocean surface and the topography of the ocean bed (ESA 1985).

## Water quality

Satellite remote sensing is good at giving information on spatial changes at the Earth's surface. One group of variables which does show suitable spatial variation is water quality. In this context water quality variables include salinity, chlorophyll *a* concentration and total suspended solids. In a series of papers Khorram (Khorram 1981, 1982, Khorram and Cheshire 1985, Cheshire *et al.* 1985) has used Landsat MSS and TM data (1) to relate *in situ* measurements of water quality to satellite radiance data, and (2) to extrapolate spatially to produce maps of water quality. The work reported by Khorram has concentrated on the San Francisco Bay delta and the Neuse river estuary, North Carolina. An example from the Neuse river estuary illustrates the general approach taken by Khorram and by other authors. For example, see Lindell *et al.* (1985) for areas off the coast of Sweden, and Carpenter and Carpenter (1983) for an example of a reservoir in Australia.

## NEUSE RIVER ESTUARY

### *Development of the models*

In the Neuse river estuary water quality samples were collected at seventy-five sites within ninety minutes of a Landsat overpass on 24 September 1982 (Khorram and Cheshire 1985). The water quality measurements were salinity, chlorophyll *a*, turbidity and total suspended solids. The Landsat MSS data for the study area were averaged for a $9 \times 9$ pixel block encompassing each water quality sample site. Each water quality variable was then related to a combination of Landsat MSS bands by multiple regression. The models were of the form

$$\hat{y} = f(L_\lambda) \tag{10.1}$$

where y is a water quality variable and $L_\lambda$ is the water radiance measured by one or more wavebands of the Landsat MSS. An example is the model for salinity (measured in parts per thousand):

$$\text{salinity} = 38.5 - 120.9 \left( \frac{\text{MSS6}}{\text{MSS4} + \text{MSS5}} \right) \tag{10.2}$$

The models for the other three variables use more complex combinations of bands, but the model form is essentially the same. Table 10.1 shows the coefficients of determination ($R^2$) at calibration for each of the four water quality models. All have respectable $R^2$ values at the model calibration stage, values which are all significant at $\alpha = 0.01$.

To verify the results the original models were applied to data for a set of

*TABLE 10.1: Coefficients of determination ($R^2$) for water quality estimates, Neuse river estuary, North Carolina. The independent variables were Landsat MSS bands in combination*

| *Water quality variable (dependent variable)* | *$R^2$ at calibration* | *$R^2$ at verification* |
|---|---|---|
| Salinity | 0.82 | 0.76 |
| Chlorophyll *a* | 0.70 | 0.48 |
| Turbidity | 0.76 | 0.72 |
| Total suspended solids | 0.64 | 0.47 |

*Source*: Khorram and Cheshire (1985)

reserved sample sites, reserved for testing purposes and not used in the development and calibration of the models. The coefficients of determination at the verification stage are also shown in Table 10.1. These $R^2$ values are relatively high for salinity and turbidity, but low for chlorophyll *a* and for total suspended solids.

The final stage in the analysis is to apply the relationships between the water quality variables and the Landsat MSS data to the entire study area. The maps produced (see colour plates 1–4 of Khorram and Cheshire 1985) show numerically the spatial distributions of the water quality variables. According to the authors:

> The distribution of these surface water quality parameters throughout the Neuse River Estuary . . . are in complete agreement with the expected and reported values of the parameters in this geographic region. (op. cit. page 337)

### *Problems*

The results described above do produce some useful maps of water quality, but there are a number of problems which the authors do not tackle and which limit the general applicability of the approach adopted.

(1) The final stage of the work is the inversion of the calibrated model, so that it implies a model of the form

$$L_\lambda = f(x) \qquad (10.3)$$

where $L_\lambda$ is again the water radiance measured at one or more of the Landsat MSS wavebands and x is a water quality variable. A model of this form is given by Aranuvachapun (1985):

$$L_\lambda = a + b(\ln S) \qquad (10.4)$$

to relate water radiance $L_\lambda$ to suspended sediment concentration S (see also Munday and Alföldi 1979). However, such inversion is not without its dangers and it cannot be assumed that a model can be inverted without modification.

(2) The models of water quality employ involved combinations of Landsat MSS bands yet the problems of multicollinearity, that is the correlation between independent variables (Johnston 1980), are not seriously addressed. For example, the turbidity model includes all four MSS bands:

$$\text{turbidity} = 0.09\left(\frac{\text{MSS4} - (\text{MSS5} \times \text{MSS7})}{\text{MSS5} - (\text{MSS6} \times \text{MSS7})}\right) - 0.43\left(\frac{\text{MSS4}}{\text{MSS6}}\right)$$
$$+21.9\left(\frac{\text{MSS4}}{\text{MSS4} + \text{MSS5} + \text{MSS6}}\right) \qquad (10.5)$$
$$-0.03\left(\frac{\text{MSS4} - (\text{MSS5} \times \text{MSS6})}{\text{MSS5} - (\text{MSS6} \times \text{MSS7})}\right) - 4.54$$

where turbidity is expressed in nephelometric turbidity units. The correlation between MSS4 and MSS5 is 0.64 and that between MSS6 and MSS7 is 0.74: these correlations are likely to bias the turbidity estimates in equation 10.5 because of the multicollinearity problem.

(3) The assessment of the accuracy of the maps is limited to the presentation of four $R^2$ values for test sites and to a qualitative assessment of the maps.

(4) The particular Landsat image chosen has cloud in its eastern part. The effects of the diffuse cloud edge and the attendant increases in atmospheric water vapour west of the cloud may have a serious impact on the measured radiances, particularly at the shorter wavelengths.

So although this form of approach is widely used in the satellite remote sensing literature, it is not without its problems. Whether for water, atmosphere or land studies inversion and multicollinearity can seriously bias the results.

## Fishing

Several of the sea-state variables discussed in this chapter have relevance to the fishing industry. The distribution of fish stocks is related partly to water temperatures, salinity and chlorophyll concentration, so an ability to map these variables operationally is of benefit to the fishing industry, provided that the information can be transmitted within hours of data capture.

It is an attractive proposition to envisage the captains of fishing vessels all over the world guiding their vessels to the best fishing grounds on the basis of the latest satellite information. Figure 10.4 shows an example of the potential of satellite remote sensing for the fishing industry off the California coast. The data on water temperature were obtained from the Nimbus 7 Coastal Zone Colour Scanner (CZCS) on 21 September 1981. The circle symbols show the location and size of albacore tuna catches during the period 19–24 September 1981. It is clear from this map that tuna can be successfully fished where coastal waters meet offshore waters, and if the CZCS data were available operationally then fishing vessels could be directed to the contact zone.

Cracknell (1983) suggests that three sources of information from satellites can assist the fishing industry: accurate weather forecasts using

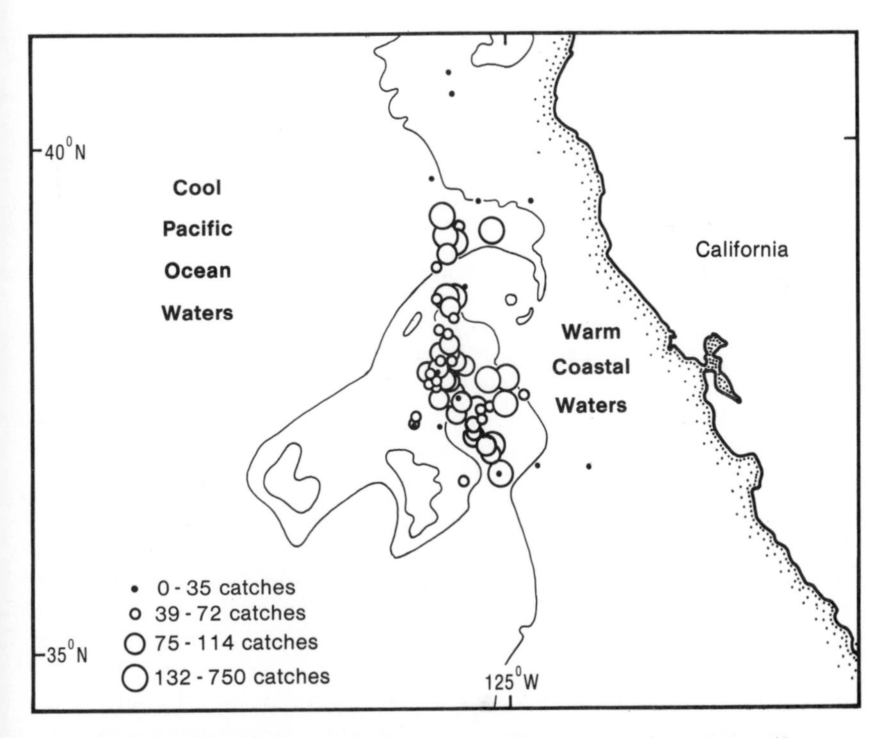


10.4 Albacore tuna catches during the period 19–24 September 1981 and water temperature chart obtained from Nimbus 7 CZCS data for 21 September 1981. (After ESA 1985)

operational meteorological satellites, sea-state data from microwave instruments such as the SMMR on Seasat and Nimbus 7, and fish location information. Table 10.2 gives a range of satellite remote sensing systems and assesses their suitability for fisheries application on a scale 0–3.

*TABLE 10.2: The suitability of satellite remote sensing systems for fisheries applications*

| | *Landsat MSS* | *Nimbus 7 CZCS* | *Noaa 5 HRIR** | *DMSP* | *HCMM* | *Seasat* |
|---|---|---|---|---|---|---|
| Temperature | 0 | 0 | 3 | 3 | 3 | 3 |
| Salinity | 0 | 0 | 0 | 0 | 0 | 1 |
| Chlorophyll | 1 | 2 | 0 | 0 | 0 | 0 |
| Colour | 2 | 3 | 0 | 0 | 0 | 0 |
| Suspended sediment | 3 | 3 | 0 | 0 | 0 | 0 |
| Sea state | 1 | 1 | 1 | 1 | 0 | 3 |
| Fronts | 2 | 2 | 3 | 3 | 2 | 2 |
| Patchiness | 2 | 1 | 1 | 1 | 1 | 2 |
| Oil | 2 | 1 | 1 | 1 | 1 | 2 |

* High Resolution Infrared

*Number code*
0 Not applicable
1 Limited value
2 Needs additional testing
3 Reliable

*Source*: Cracknell (1983), after Klemas (1981)

Lasker *et al.* (1981) give an example of the use of Noaa 6 AVHRR thermal infrared data for analysing fish distributions. They concentrated on the coastal waters off the southern California coast and found that anchovies at their peak of spawning avoided a water mass with a temperature $\leqslant 14°$ C which had entrained into the California Current from cold, upwelling water further north. By contrast, warmer water pools at 16° C, such as those in the wake of Santa Catalina Island, provided good conditions for anchovy spawning. In this example the Noaa AVHRR data assisted the oceanographers in determining the coastal water dynamics and so the likely behaviour of the anchovies, rather than simply assisting the mapping of the surface water temperature distribution.

In 1981 an experimental programme in the eastern Pacific Ocean to assess the role of satellite data for commercial fisheries was jointly sponsored by NASA and the Jet Propulsion Laboratory, with the cooperation of a number of other agencies including the Ocean Services Division of the US National

Weather Service. The experiment used Noaa AVHRR, Nimbus 7 CZCS and GOES data together with conventional observations used by the US Navy. Operational charts were produced from this range of data and transmitted by radio facsimile to ships in the eastern Pacific. The operational charts included:

(1) Sea surface temperature.
(2) Mixed layer depth.
(3) Wind direction and speed.
(4) Wave height and direction.
(5) Ocean colour zones and boundaries.

This information was used primarily to plot the likely location of three fish species: salmon, albacore tuna and tropical tuna. Operational maps were produced regularly from 1981 to 1983 showing the distribution of these species, and Cracknell (1983) gives examples of the maps. The conclusion of this experiment was that satellite remote sensing data can be a cost-effective tool in assisting the fishing industry operationally.

## Surface biology

At shorter wavelengths the variations in water colour can be identified from satellite platforms, and these variations can be related to surface water composition and biological activity. The CZCS carried on Nimbus 7 was designed for this task. It had six narrow wavebands, four of which were related to the spectral characteristics of phytoplankton. The CZCS imagery can provide information on the concentration of marine pigments (chlorophyll *a* and detrital phaeopigments) and from this it is possible to infer the amount of phytoplankton biomass. Phytoplankton concentration C can be estimated by

$$C = ar_{ij}{}^{b} \qquad (10.6)$$

where a and b are constants and $r_{ij}$ is the ratio of water-leaving radiance in two wavebands i and j (Aranuvachapun 1983). For low phytoplankton concentrations Caraux and Austin (1983) used CZCS bands 1 and 3, centred at 0.443μm and 0.55μm respectively, as i and j in equation 10.6 and for high concentrations of phytoplankton they used bands 2 and 3, 0.52μm and 0.55μm, for i and j respectively. The two equations used by Caraux and Austin (1983) in a study of the Gulf of Lyons in the western Mediterranean Sea were:

(1) For low concentrations of phytoplankton, $C \leqslant 1.5\text{mg m}^{-3}$

$$C = 1.129 \left[\frac{L_{0.443\mu m}}{L_{0.55\mu m}}\right]^{-1.711} \quad (10.7)$$

(2) For high concentrations of phytoplankton, $C > 1.5\text{mg m}^{-3}$

$$C = 3.326 \left[\frac{L_{0.52\mu m}}{L_{0.55\mu m}}\right]^{-2.439} \quad (10.8)$$

where L is the water-leaving radiance in the CZCS waveband specified by the subscript, suitably corrected for atmospheric effects (Gordon *et al.* 1980, Robinson 1983).

The phytoplankton maps of the Gulf of Lyons prepared by Caraux and Austin (1983) reveal two features. Firstly, a mesoscale cyclonic eddy can be observed off the coast south-west of Marseilles. The eddy is some 50km in diameter and is an area of low phytoplankton concentration. Secondly, the outflow from the River Rhone into the Mediterranean Sea produces a plume of high phytoplankton concentration which has a north-east–south-west orientation, and on occasions of strong winds, such as during a Mistral, the plume can reach the Spanish coast some 200km away.

Eddy and plume features similar to these but covering larger areas can also be seen in CZCS data of the North Atlantic (ESA 1985). High concentrations of phytoplankton are found inshore near the eastern coast of the USA where nutrients are available from river runoff, and cool, nutrient-rich deep water is mixed by tides and winds. The Gulf Stream itself is relatively poor in nutrients, and the eddies which break away to the north of the Gulf Stream (see Figure 10.2) are circulations of low phytoplankton concentration.

Eddy structures can also be seen in Landsat MSS imagery of water bodies. Ulbricht (1983a, b) gives an illustration of a 1980 Landsat MSS image of the south-western Baltic Sea which shows accumulations of blue–green algae or cyanobacteria in fine eddy and filament structures. The eddies themselves have their origin in the bottom topography of the Baltic Sea. The algae bloom in the Baltic during July, August and September and can fix nitrogen directly from the atmosphere. Their gas vacuoles which enable this fixation make the algae float at or near the surface and so become visible, particularly at shorter wavelengths such as Landsat MSS band 4. A quantitative estimate of nitrogen fixation can be achieved by measuring the spatial extent of these algal blooms.

Finally, data from the Nimbus 7 CZCS have been used in a discovery role. In the summer months patches of highly reflective water some $100{,}000\text{km}^2$ in extent show up on CZCS imagery of the North Atlantic west of the

British Isles (NERC 1985). These highly reflective patches are caused by coccolithophores, a group of phytoplankton. The coccolithophores contribute significantly to the ocean fluxes of carbon and are important in the transfer of carbon dioxide from the atmosphere to the oceans. Their changing pattern over time discerned from satellite data can be related to ocean sediment data so that the dynamics of this organism can be analysed.

## Snow

Research into ice and snow has used a wide variety of satellite data. Microwave wavelengths, using both passive and active sensors, have been particularly useful as they can be used in cloudy conditions. Seasat SAR imagery has been used to map changes in sea ice patterns, and Seasat altimeter data have been used to determine changes in the elevation of ice sheets to better than ±1m accuracy and to sound ice thickness (Robin *et al.* 1983). Landsat MSS and TM data have been used to map the spatial extent of ice and snow in cloud-free areas, and if sufficient imagery can be obtained over time then attempts at mapping snow accumulation and snow melt can be made. NOAA maps snow cover extent from visible and infrared satellite remote sensing data on a routine basis for several river basins in the western United States (Burke *et al.* 1984).

One sensor which has been particularly useful for monitoring snow is the Electrically Scanning Microwave Radiometer (ESMR) which was carried on the Nimbus 5 and Nimbus 6 satellites. The ESMR operates in the microwave region of the EM spectrum: the ESMR on Nimbus 5 (ESMR-5) at 1.55cm wavelength and the ESMR on Nimbus 6 (ESMR-6) at 0.81cm wavelength. The radiation measured by the ESMR, commonly termed the brightness temperature $T_B$, is a function of both the emissivity and the temperature of a snowpack. The emissivity can be influenced by a number of snow properties including snow water equivalent, density, liquid water, and grain and crystal sizes.

One snow variable of considerable practical and scientific importance is snow depth. Foster *et al.* (1980) explored the relationships between snow depth and ESMR $T_B$ measurements for three large and comparable geographical areas: the high plains of Montana and North Dakota in the USA, the high plains of Canada and the steppes of central Russia. Figure 10.5 shows the composite scatter plot of the relationship of snow depth with ESMR-6 brightness temperature for data from early 1976. The regression relationship is

$$s = 46.99 - 0.178T_B \tag{10.9}$$

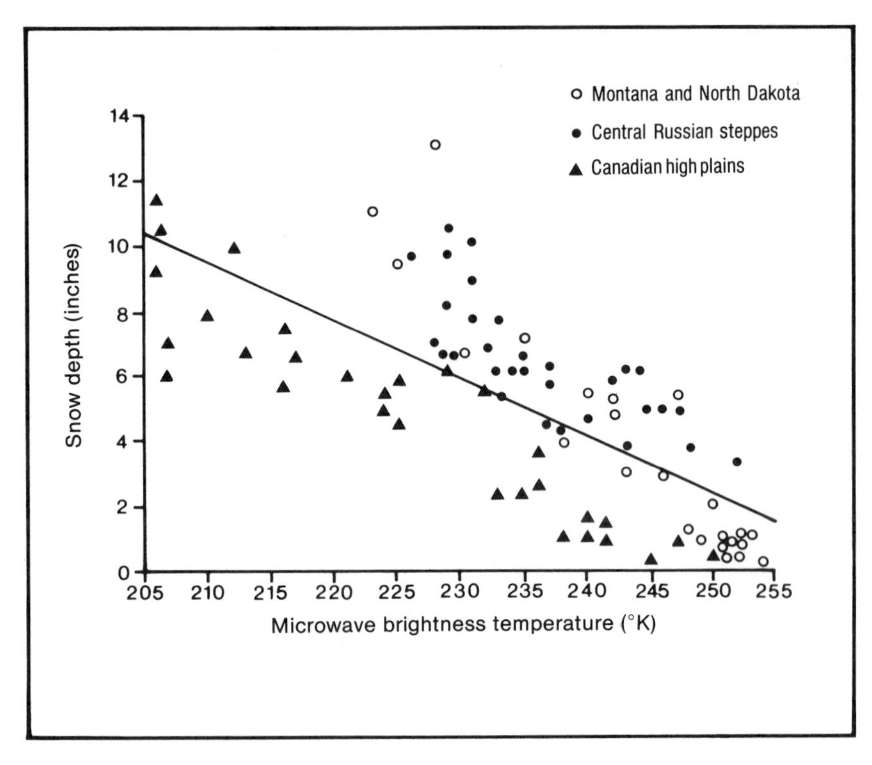


10.5 Relationship between snow depth and Nimbus 6 ESMR brightness temperature data for three areas in North America and Russia. (Modified from Foster *et al.* 1980)

where s is snow depth in inches and $T_B$ is ESMR-6 brightness temperature in deg K. The correlation between the two variables is 0.71 and is significant at $\alpha = 0.01$. However, the scatter of points about the regression line is rather large and there appears to be some patterning in the scatter plot: the data from the Canadian high plains are mostly below the regression line, while the data from Russian steppes are generally above the regression line. This suggests a geographical effect which may be connected with elevation or with terrain cover below the snowpack.

Foster *et al.* (1980) analysed the snow depth and brightness temperature relationships for separate geographical areas and the results are shown in Table 10.3. In all cases the results from the shorter wavelength ESMR sensor on Nimbus 6 are better than those from the ESMR on Nimbus 5. The shorter wavelength 0.81cm data appear to be more sensitive to the structure and condition of the snow than the 1.55cm data which are additionally affected by underlying soil conditions.

The use of Nimbus ESMR data for snowpack measurement was extended

*TABLE 10.3: Correlations of snow depth with brightness temperature for three geographical areas*

| *Area* | *Nimbus 5 ESMR (1.55cm wavelength)* | *Nimbus 6 ESMR (0.81cm wavelength)* |
|---|---|---|
| High plains of Montana and North Dakota, USA | 0.90 | 0.94 |
| High plains of Canada | 0.87 | 0.93 |
| Russian steppes | 0.72 | 0.77 |

*Source*: Foster *et al.* (1980)

by Burke *et al.* (1984) who used ESMR-5 and ESMR-6 data together with data from ESMR's successor, the Scanning Multichannel Microwave Radiometer (SMMR) carried on Nimbus 7. Burke *et al.* again concentrate on the relationship between snow and microwave brightness temperature, and reach a number of conclusions:

(1) Snow boundaries can usually be defined because of the sharp decrease in $T_B$ from land to snow.
(2) For dry snow conditions the 0.81cm wavelength data display a decrease in $T_B$ as a function of snow depth because of the stronger volume scattering effect of deeper snow. At 1.7cm wavelength the sensitivity of $T_B$ to snow depth is not significant.
(3) The onset of snowmelt can be determined at both 0.81cm and 1.7cm wavelengths because of the significant increase in $T_B$.
(4) On balance the 0.81cm data are preferred to the longer wavelength data for snow studies, as Foster *et al.* (1980) found previously.

These conclusions indicate that passive microwave measurements are suitable for snowpack monitoring. Schanda *et al.* (1983) illustrate this by showing snow cover maps for the whole northern hemisphere obtained by processing Nimbus 7 SMMR data for 1979. They were able to categorise snow deeper than 10cm, thin snow of less than 10cm, and snow-free areas of the northern hemisphere by using the differences between the 0.81cm and 1.7cm SMMR brightness temperature data.

*Chapter 11*

# Conclusion: the way forward

The satellite remote sensing scene is in a continual state of flux. New satellites are being launched and new ideas for satellite sensors are emerging at a rapid rate. It will be decades yet before any satellite observing system for the Earth settles down into a stable pattern. With this in mind this concluding chapter will not attempt to cover every planned satellite system for remote sensing, but will extract some of the main themes from known developments in this fast-moving field. The first part of the chapter discusses planned and approved satellite systems, the second part discusses the plans for a space station and the final part of the chapter discusses NASA's proposals for a broad-based Earth Observing System.

## International coordination

In September 1984 in Washington DC an informal grouping of countries and agencies met to discuss international coordination of satellite remote sensing programmes. The group called itself the Committee for Earth Observation Satellites (CEOS) and it is a committee which is concerned with Earth observation satellite systems, excluding meteorological satellites which already have appropriate forums for discussion: the committee for the Coordination of Geostationary Meteorological Satellites (CGMS) and the committee for International Polar Orbiting Meteorological Satellites (IPOMS) (ESA 1984).

CEOS acts as a forum for the exchange of technical information to encourage complementarity and compatibility among space-borne Earth observing systems, and focuses on those systems which have been planned and approved. A list of such satellite systems considered by CEOS is given in

Table 11.1. One element which is clear from Table 11.1 is that the near monopoly enjoyed by the United States in Earth observation is nearing an end (ESA 1984), and although the United States still retains a very strong satellite programme in this area, there is competition from Brazil, Canada, China, Europe, India and Japan.

*TABLE 11.1: Approved and planned polar-orbiting Earth observation satellite systems*

| | *Mission* | *Objectives* | *Main sensors* | *Altitude (km)* | *Launch date* | *Status** |
|---|---|---|---|---|---|---|
| zil | Remote Sensing Satellite | Land applications | Multispectral push-broom imager | 650 | 1990/91 | Phase B |
| nada | Radarsat | Ice and land monitoring | Synthetic aperture radar<br>Wind scatterometer<br>Optical sensor | 1000 | 1991 | Phase B |
| ina | Earth resources satellite | Land applications | Multispectral push-broom imager | 400 | 1988/89 | Phase B |
| rope A) | ERS-1** | Ocean-ice monitoring | Synthetic aperture radar<br>Wind scatterometer<br>Radar altimeter<br>Along-track scanning radiometer<br>Microwave sounder<br>Precise positioning package | 780 | 1989/90 | Phase C/D |
| nce | SPOT 2 | Land applications | High resolution visible imager (HRV) | 832 | 1987 | A |
| | SPOT 3 and 4 | Land applications | Improved HRV<br>Medium resolution sensor | | 1990 and 1994 | |
| | Poseidon (on SPOT 3) | Ocean circulation | Radar altimeter<br>Microwave sounder<br>Precise positioning system | 832 | 1990 | Phase B |
| ia | IRS-1 | Land applications | LISS I and II | 904 | 1987 | A |
| an | MOS-1** | Marine applications | Multispectral electronic self-scanning radiometer<br>Visible and thermal infrared radiometer<br>Microwave scanning radiometer | 909 | 1987 | A |
| | J-ERS-1 | Land applications | Synthetic aperture radar<br>Visible and near infrared radiometer | 570 | 1990/91 | A |

*Table 11.1 cont.*

| | *Mission* | *Objectives* | *Main sensors* | *Altitude* | *Launch* | *Stat* |
|---|---|---|---|---|---|---|
| *United States* | | | | | | |
| NASA + France | TOPEX Poseidon | Ocean circulation | Radar altimeter<br>Microwave radiometer<br>Microwave sounder<br>Precise positioning system | 1300 | 1990/91 | Pha B |
| US Navy + NASA + NOAA | N-ROSS** | Ocean applications | Radar altimeter<br>Wind scatterometer<br>Microwave imager<br>Microwave radiometer | 830 | 1990/91 | A |
| NOAA | Advanced Tiros-N series (NOAA K,L, M,N,O,P) | Weather monitoring | AVHRR<br>Microwave sounding unit<br>High resolution infrared sounder | c.850 | 1989 onwards | O |
| NOAA + Eosat | Landsat 6 and 7 | Land applications | Thematic mapper<br>Multilinear array | 700 | 1993 | A |
| US Navy | DMSP series | Weather monitoring | Linescanner<br>Microwave imager<br>Infrared and micro-wave sounders | 833 | 1987 onwards | O |

*Source*: ESA (1986)

* The status categories are:

| | |
|---|---|
| A | Approved |
| Phase B | Detailed definition phase |
| Phase C/D | Development, manufacture and integration phase |
| O | Operational |

** There are plans for follow-on satellites in these missions

It is not the intention here to discuss all the satellites and sensors in Table 11.1, but rather the discussion will focus on two systems which are in many ways representative of a variety of satellites: Landsat and ERS-1. Landsat provides a link with the progress to date in visible and infrared satellite remote sensing. ERS-1 is the first operational European remote sensing satellite and is equipped with microwave sensors for a variety of applications; this provides a link with many future satellites which will employ passive and active microwave technology.

## LANDSAT

### *Timetable*

The Landsat 5 satellite has a design lifetime of three years. If it fails on schedule at the end of its design life in February 1987 then there will be no Landsat satellite in full operation for some years. Landsat 6 is planned for launch in 1993, so there is a potential data gap from the end of Landsat 5 operations to the commencement of Landsat 6 of up to six years (NOAA 1986, *Flight International* 18 October 1986). These next two Landsat satellites will be orbited on a new Omnistar platform or bus, which is capable of taking heavier payloads and can be serviced in orbit. However, the Omnistar bus is designed for launch from the Space Shuttle, and following the Challenger accident of 1986 the Shuttle programme has been significantly delayed. The number of Space Shuttles is down from four to three, although NASA has been allowed to order a replacement for Challenger, the restart of Shuttle flights is delayed until 1988, and when the Shuttle service does start again the military missions will dominate the launch schedule in the interests of US national security.

### *Payload*

The Landsat 6 payload will be an extension of the payload on Landsats 4 and 5. An Enhanced Thematic Mapper (ETM) and an Emulated Multi-spectral Scanner (EMSS) are planned. The ETM will have the same bands as the earlier Thematic Mapper (TM) with 30m pixels (see Table 5.7 in chapter 5), but will also carry a new panchromatic band with 15m pixels and a wavelength range of 0.5–0.8μm. All the ETM channels will not be available simultaneously. For example, the panchromatic channel will be available with a selected number of the TM channels, or all TM channels will be available but without the panchromatic channel. The EMSS instrument will use the ETM channels 1–4 to emulate the earlier MSS carried on all previous Landsat satellites. The EMSS channels will have the same wavebands as TM 1–4, but will have a 60m pixel size by processing the ETM data on board the satellite.

Landsat 7 will carry an ETM as planned for Landsat 6, but instead of just one thermal infrared band there will be up to four bands in the wavelength range 8–11.6μm with a pixel size of 60m. Landsat 7 will also carry an EMSS for data continuity and a multi-linear array is a possible component of the payload. As with Landsat 6 there will be a limit to the number of channels available because of data rate transmission constraints, and not all channels will be available simultaneously.

### *Competition*

While the plans for Landsat are definite enough, and the operator Eosat is committed to them, the future of the Landsat programme does not seem as rosy as its past. The competition described in Table 11.1 is arriving at a time when the successful sequence of Landsat satellites is suffering problems, and in retrospect Landsat may be seen as a long-lived experimental programme which developed the market which operators of other satellite systems such as SPOT can now exploit.

## ERS-1

Europe's first remote sensing, rather than meteorological, satellite is ERS-1 which is planned for launch in late 1989 or 1990. A diagram illustrating ERS-1 is shown in Figure 11.1. ERS-1 will be an operational satellite providing a regular and reliable supply of data to a wide variety of users

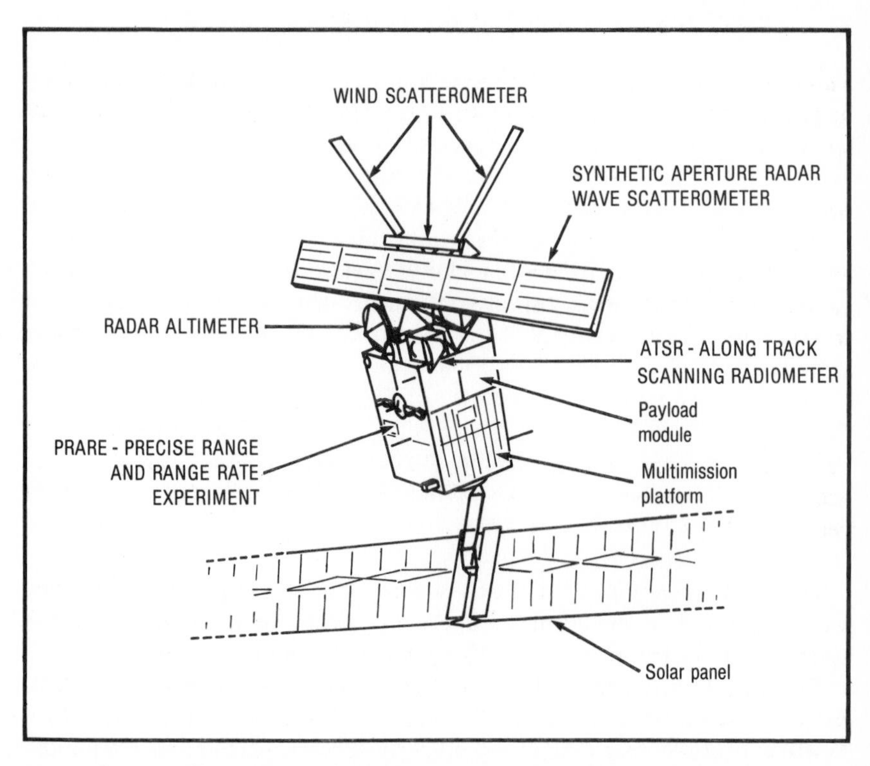


11.1 Schematic illustration of the ERS-1 satellite. (Modified from NRSC 1985)

including shipping and fishing companies, and the oil industry. The objectives of ERS-1 are primarily intended to enable the exploitation of coastal oceans and to improve global meteorological information by collecting data over the world's oceans. In addition, the synthetic aperture radar carried on ERS-1 will be a benefit to land applications of satellite remote sensing.

Four main instruments will be carried as the sensor payload on ERS-1 and these instruments are listed below (Haskell 1983).

*Active Microwave Instrument (AMI)*. The AMI will have three switchable modes:

(1) Synthetic aperture radar (SAR) mode. The SAR is a C-band (c. 5cm wavelength) instrument with a 30m pixel size and an 80km swath width. The SAR has, of course, an all-weather capability.
(2) Wave mode. The height and wavelength of ocean waves will be calculated from their spectra for sample areas of 5km × 5km.
(3) Wind mode. Wind speed and direction at the ocean surface will be measured for areas 25km × 25km in size.

*Radar Altimeter (RA)*. The altimeter will be used to measure ocean wave heights in the range 1–20m with an accuracy of 0.1–0.5m, and to measure ice topography with a height accuracy of 0.4m.

*Along-Track Scanning Radiometer (ATSR-M)*. This instrument will measure sea surface temperatures to an absolute accuracy of better than ± 0.5 deg K, and will collect observations of clouds, aerosols, haze and total water vapour content of the atmosphere. The instrument will have a microwave sounder and channels at 3.7, 11 and 12μm, and will have a pixel size of 1km × 1km.

*Precise Range and Range-Rate Experiment (PRARE)*. This instrument will be an experiment in determining the satellite's orbital altitude to an accuracy of approximately 10cm. This information will then be used for geodesy and geodynamics studies.

The data from ERS-1 will be used in a wide range of applications, including scientific, experimental and commercial areas. Table 11.2 gives a match of the instruments to the likely applications of the data.

Table 11.1 shows that ERS-1 will carry a payload which has much in common with other planned satellite systems. Radarsat, Poseidon, J-ERS-1, TOPEX, N-ROSS and DMSP will all carry sensors which have comparable characteristics to ERS-1, and it is clear that for many future missions

*TABLE 11.2: Relationships of instruments to applications of ERS-1 data*

| | ERS-1 INSTRUMENT | | | | |
|---|---|---|---|---|---|
| | | | Active Microwave Instrument | | |
| *Application* | *Radar altimeter and PRARE* | *ATSR-M* | *Imaging mode (SAR)* | *Wave mode* | *Wind mode* |
| Weather forecast | | x | | | x |
| Sea-state forecast | x | | | x | x |
| Offshore activity | x | x | | x | x |
| Ship routing | x | x | | x | x |
| Fish location | x | x | | | x |
| Iceberg monitoring | (x) | (x) | x | | |
| Oil and pollution detection | | x | x | | |
| Coastal processes | | x | x | | |
| Land applications | (x) | | x | | |
| Ocean circulation | x | x | (x) | | x |
| Ocean tides | x | | | | |
| Wind fields | x | | (x) | x | x |
| Wave fields | x | | (x) | x | x |
| Polar oceans | x | x | x | x | x |
| Land ice | x | | (x) | | |
| Sea surface temperature | | x | | | |
| Marine biology | | (x) | | | |

*Source*: NRSC (1985)

all-weather information for land, sea and ice areas will be vital to the exploitation of satellite remote sensing. This is not to deny the importance of visible and infrared imaging of the Earth's surface, and this will be achieved by satellites such as Landsat, SPOT, MOS-1, IRS-1, Tiros-N and DMSP.

## EUROPEAN PROGRAMME

Through a variety of mechanisms European countries are collaborating in satellite remote sensing. The EUMETSAT organisation has been established to operate the series of Meteosat satellites which constitute the Meteosat Operational Programme: three additional satellites will be launched in

1988, 1989 and 1990 to provide geostationary meteorological satellite data already so well established by the first two Meteosat satellites. European collaboration is considerably larger than this and the European programme in Earth observation has four key mission areas in a complementary strategy (ESA 1985).

| | |
|---|---|
| *Atmosphere* | Development of a second generation Meteosat. |
| *Oceans and ice* | Continuation of the ERS-1 mission through a second flight model, ERS-2.<br>Development of a prototype advanced satellite for operational ocean and ice monitoring. |
| *Land* | Development of an advanced optical and all-weather microwave satellite for land applications and sciences. |
| *Solid earth* | Development of a precise point positioning geodetic satellite for Earth dynamics research. |

Some elements of this European strategy already have firm plans while others are at an early stage of development. Figure 11.2 shows the timeline for the proposed European programme in Earth observation, which in 1990 would cost approximately 200 million accounting units annually (1984 prices).

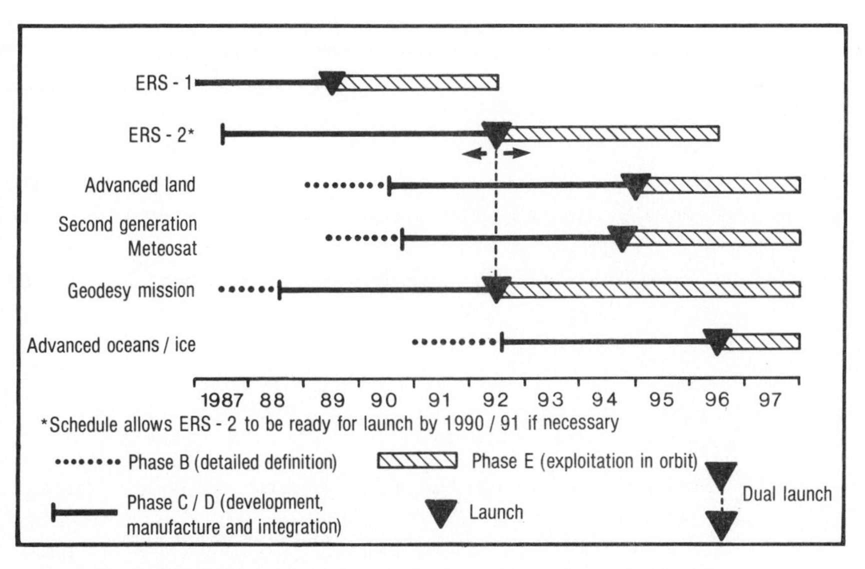


11.2 A European programme in Earth observation. (After ESA 1985)

## Space station and Columbus

### THE CONCEPT

The list of satellite systems given in Table 11.1 shows the approved and planned satellites over the next few years. The next scene to be acted out in the drama of satellite remote sensing will take place on a wider stage. Sometime in the 1990s it is likely that an international space station will be placed into orbit around this small planet and will provide a focus for a large number of space-borne experiments. In this context satellite remote sensing will be just one part of a wider activity to exploit space for scientific and commercial purposes. The plan for a space station, a permanent presence in space, was initiated by the United States and has strong backing from Europe, Canada and Japan. In his 1984 State of the Union address President Reagan directed NASA 'to develop a permanently manned space station and to do it within a decade'. This initial proposal was to construct a space station between 1993 and 1995 using a series of Shuttle flights to take the components into orbit (Furniss 1986).

The central part of the proposed space station is a manned station in a circular orbit at an altitude of approximately 500km and an inclination angle of 28.5° (Lodge 1985). The manned station will have living and working quarters for astronauts, and will be a centre for communications, assembly, transport and servicing.

For remote sensing purposes an orbit inclination angle of 28.5° has a limited value as only the equatorial and sub-tropical regions can be covered. To provide global coverage two polar orbiting platforms are proposed which will carry a variety of remote sensing sensors. Europe is particularly interested in a polar orbiting platform and this is one part of the European effort termed the Columbus programme. Columbus has a number of elements which contribute to the space station concept, including a pressurised module, a resource module, and man-tended platforms.

In many ways the forces behind Columbus and the space station have not been dominated by scientific enquiry, but by a political desire to exploit space which some argue is the final frontier. Indeed, Lodge (1985) has commented that

> the political, technological and industrial momentum is very probably so great that the luxury of making a decision (on the space station concept) on scientific and objective grounds as to whether this is the right way to go cannot be entertained.

One important feature of the space station is that it will be visited for construction and servicing by the Space Shuttle. Clearly the Challenger accident in 1986 has had an impact on the planning and sequencing of Shuttle flights and there will certainly be problems in building a space station by the mid-1990s as was originally proposed. The European Ariane rocket does provide an alternative launch vehicle, although this too is not without its problems as failed launches in 1985 and 1986 have shown.

## POLAR ORBITING PLATFORM

As indicated above the polar orbiting platform elements of the space station will provide the most suitable opportunities for remote sensing. Two polar platforms are proposed: one to be provided by ESA and placed in an orbit to give a morning equatorial crossing time at around 10 a.m.; the second to be provided by NASA and placed in a similar orbit but with an afternoon equatorial crossing time at around 2 p.m. The preferred orbit is sun-synchronous at c. 850km altitude. Each platform will have a payload limit of approximately 2000 kg, and will have a servicing interval of two to three years.

The list of candidate instruments for the polar platform is already large. In Table 11.3 a list is given of candidate instruments suggested by the Anglo–French Columbus utilisation working group in February 1986. This list has been quickly overtaken by events but it does give an indication in detail of the type of proposals under review. The list in Table 11.3 is already a short list taken from a longer list of desirable instruments, but even Table 11.3 is too long for the weight, size and power constraints of the polar platform. Different combinations of a selection of about a dozen of these instruments with a total mass of c. 2000 kg have been evaluated by ESA (1986), bearing in mind the needs of users in different disciplines and the operational, experimental and demonstration requirements of the polar platform.

*TABLE 11.3: Candidate instruments suggested for the polar platform by the Anglo–French Columbus utilisation working group, February 1986*

*Meteorological observations*

Advanced Very High Resolution Radiometer
High Resolution Infrared Radiation Sounder
Advanced Microwave Sounding Unit-A
Advanced Microwave Sounding Unit-B
Single Frequency Atmospheric Lidar

*Atmospheric observations*

Microwave Limb Sounder
Improved Stratospheric and Mesospheric Sounder

*Atmospheric observations (cont.)*

Doppler Wind Sounder
Cooled Infrared Spectrometer
$HO_x$ monitor
Global Ozone Monitoring Radiometer
Earth Radiation Budget sensor
Conically Scanning Radiometer

*Ocean/ice observations*

Ocean Temperature Large Antenna Microwave Radiometer
Multifrequency Microwave Radiometer
Advanced Radar Altimeter
Microwave Wind Scatterometer
Ocean Colour Monitor
Along Track Scanning Radiometer
Ocean-Ice-Land Synthetic Aperture Radar*
Lidar Altimeter/Ranger

*Land observations*

Thematic Mapper Derivative
Synthetic Aperture Radar*
Corner Cube Reflectors
Gravitometer
Stereoscopic High Resolution Imager
Geophysics Magnetometer

*Astronomy*

Solar Astronomy Package
Dust Environment Monitor
High Energy Astronomy Package

*Solar terrestrial physics*

Solar Proton Monitor
Solar Input Monitor
Passive Magnetospheric Plasma Probe
Auroral Imager

*Support*

Data Collection and Location System
Precise Positioning System
Search and Rescue Satellite System
Biological Environment Monitor
System Environment Monitor
Transmitter Package
Data Storage System

*Source*: Anglo–French Columbus utilisation working group
* Same instrument

From a European viewpoint it is not clear how the satellites listed in Table 11.1 will interact with the polar platform. Will satellites such as ERS-1, SPOT and Radarsat all be independent programmes or will their sensors be incorporated in some way in the polar platform? Can the polar platform be successful if it is in a sense in competition with the operators of independent Earth observation satellites?

## Earth Observing System

The discussion above has rather concentrated on the European dimension to a polar platform. In the United States a study team in NASA began work in 1981 to examine the broad issues facing the Earth sciences and how best they can be analysed. The result of this study team's work is a report on an Earth Observing System (EOS) (NASA 1984), which is one view of the characteristics of a space platform in low Earth orbit. The study team's report states clearly that 'Earth science research is now ready for a unified approach based upon the view that the physical, chemical and biological processes at work on Earth comprise a coupled global system'.

The EOS report considers the goals for the Earth sciences in the 1990s under four headings: the hydrological cycle, biogeochemical cycles, climatological processes and geophysical processes including those in the atmosphere, the oceans and the solid Earth. From these goals a number of EOS instruments have been defined which will allow study of the integrated nature of global environmental processes and which could contribute to a polar platform. The list of instruments, thirteen in the EOS report, has similarities to those in Table 11.3, except that it is assumed that the Landsat, Noaa and GOES satellite programmes will continue independently of the polar platforms. The five recommendations of the EOS report have implications which go beyond those for a space station and polar platform, and are included below as they can be seen as providing some signposts for satellite remote sensing in the longer term.

(1) A program must be initiated to ensure that present time series of Earth science data are maintained and continued. Collection of new data sets should be initiated.
(2) A data system that provides easy, integrated, and complete access to past, present, and future data must be developed as soon as possible.
(3) A long-term research effort must be sustained to study and understand these time series of Earth observations.
(4) The Earth Observing System should be established as an information

system to carry out those aspects of the above recommendations which go beyond existing and currently planned activities.

(5) The scientific direction of the Earth Observing System should be established and continued through an international steering committee.

Environmental systems do not have national boundaries, and so international efforts at analysing coupled global systems from the unique viewpoint of space are clearly needed. There is evidence that some of the issues listed above are being approached, for example with the establishment of the International Satellite Land Surface Climatology Project, and indeed the space station concept embodies a number of these recommendations.

One important component of the EOS recommendations is the need for a data system. This will need to go beyond the scope and scale of existing Geographical Information Systems (Jackson and Mason 1986) and provide rapid access to very large volumes of information which are continually being updated, as well as providing access to time series of historical data.

By the end of the twentieth century a large proportion of the observation needs of the Earth sciences will be capable of being met by satellite observations, combined where necessary with surface observations. Satellite remote sensing is now providing a global coverage of environmental data for this small planet, a coverage which is now the largest data set of environmental information available to the scientific community. Gregory (1985) suggests that remote sensing 'is now poised to offer even greater insights into the environment'. This poise must be converted into action in the future as environmental scientists will increasingly turn first to the largest quantity of data for the study of the Earth – the data provided by satellite remote sensing.

*Appendix 1*

# Abbreviations and acronyms

| | |
|---|---|
| AMI | Active Microwave Instrument, on ERS-1. |
| APT | Automatic Picture Transmission. |
| ATI | Apparent thermal inertia. |
| ATS | Applications Technology Satellite. |
| ATSR-M | Along-Track Scanning Radiometer, on ERS-1. |
| AVHRR | Advanced Very High Resolution Radiometer, on Noaa satellites. |
| CCD | Charge-coupled device. |
| CCT | Computer-compatible tape. |
| CEOS | Committee for Earth Observation Satellites. |
| CGMS | Coordination of Geostationary Meteorological Satellites committee. |
| CRC | Colour ratio composite. |
| CZCS | Coastal Zone Colour Scanner, on the Nimbus 7 satellite. |
| DMSP | Defense Meteorological Satellite Program. |
| DoE | Department of the Environment. |
| ECMWF | European Centre for Medium Range Weather Forecasting. |
| EM | Electromagnetic. |
| EMSS | Emulated Multispectral Scanner, on Landsats 6 and 7. |
| EOS | Earth Observing System. |
| Eosat | Earth Observation Satellite Corporation. |
| ERB | Earth Radiation Budget experiment, carried on the Nimbus 7 satellite. |
| ERS-1 | European remote sensing satellite. |
| ERTS | Earth Resources Technology Satellite, renamed Landsat. |
| ESA | European Space Agency. |
| ESMR | Electrically Scanning Microwave Radiometer, carried on Nimbus satellites. |
| ESOC | European Space Operations Centre. |
| ESSA | Environmental Science Services Administration. |
| Essa | ESSA satellite. |
| ETM | Enhanced Thematic Mapper. |
| EUMETSAT | European Meteorological Satellite organisation. |
| FGGE | First GARP Global Experiment. |

| | |
|---|---|
| FIFE | First ISLSCP Field Experiment. |
| GAC | Global area coverage, data resampled from Noaa AVHRR data. |
| GARP | Global Atmospheric Research Programme. |
| GCP | Ground control point. |
| GEO | Geostationary Earth orbit. |
| GLAS | Goddard Laboratory for Atmospheric Sciences. |
| GMS | Geostationary Meteorological Satellite. |
| GMT | Greenwich Mean Time. |
| GOES | Geostationary Operational Environmental Satellite. |
| GVI | Global Vegetation Index. |
| HCMM | Heat Capacity Mapping Mission. |
| HCMR | Heat Capacity Mapping Radiometer, on HCMM. |
| HIRS2 | High Resolution Infrared Sounder, on Tiros-N/Noaa satellites. |
| HRPT | High Resolution Picture Transmission. |
| HRV | High Resolution Visible, sensor on the SPOT satellite. |
| IFOV | Instantaneous field of view. |
| IPOMS | International Polar Orbiting Meteorological Satellites committee. |
| IRS-1 | Indian remote sensing satellite. |
| ISLSCP | International Satellite Land Surface Climatology Project. |
| ITOS | Improved Tiros Operational System. |
| J-ERS-1 | Japanese environmental remote sensing satellite. |
| LAC | Local area coverage, data from Noaa AVHRR. |
| LACIE | Large Area Crop Inventory Experiment. |
| LEO | Low Earth Orbit. |
| LFC | Large Format Camera, on the Space Shuttle. |
| MLA | Multi-linear Array. |
| MOMS | Modular Opto-electronic Multispectral Scanner. |
| MSS | Multispectral Scanner, on Landsats 1–5. |
| MSU | Microwave Sounding Unit, on Tiros-N satellites. |
| NASA | National Aeronautics and Space Administration. |
| NOAA | National Oceanic and Atmospheric Administration. |
| Noaa | NOAA satellite. |
| NRSC | National Remote Sensing Centre. |
| NVI | Normalised Vegetation Index. |
| OCRS | Ontario Centre for Remote Sensing. |
| PCA | Principal components analysis. |
| PRARE | Precise Range and Range-Rate Experiment, on ERS-1. |
| PVI | Perpendicular Vegetation Index. |
| RBV | Return Beam Vidicon, on Landsats 1–3. |
| SAR | Synthetic aperture radar. |
| SIR | Shuttle Imaging Radar. |
| SLAR | Side-looking airborne radar. |
| SMS | Synchronous Meteorological Satellite. |
| SMSA | Standard Metropolitan Statistical Area. |
| SOM | Space Oblique Mercator projection. |
| SPOT | Satellite Probatoire d'Observation de la Terre. |
| SSM/I | Special Sensor Microwave/Imager, on DMSP satellites. |
| SSU | Stratospheric Sounding Unit. |
| TDRSS | Tracking and Data Relay Satellite System. |

| | |
|---|---|
| Tiros | Television and Infrared Observation Satellite. |
| TM | Thematic Mapper, on Landsats 4 and 5. |
| TOVS | Tiros Operational Vertical Sounder. |
| USDA | United States Department of Agriculture. |
| USGS | United States Geological Survey. |
| UT | Universal Time. |
| UTM | Universal Transverse Mercator projection. |
| VISSR | Visible and Infrared Spin-Scan Radiometer, on SMS satellites. |
| WMO | World Meteorological Organisation. |
| XS | Multispectral channel on the SPOT HRV. |

*Appendix 2*

# International Landsat distribution centres

*(1) United States of America*

Eosat
c/o Landsat Customer Services
EROS Data Center
Sioux Falls, SD 57198
Telephone: (605) 594-2291 or 1-800-367-2801
TWX: 910-668-0310-EDC SFL

*(2) Brazil*

Instituto de Pesquisas Espaciais (INPE)
Departamento de Producao de Imagens
ATUS-Banco de Imagens Terrestres
Rodovia Presidente Dutra, Km 210
Cachoeira Paulista-CEP 12.630
Sao Paulo, Brazil
Telephone: (0125) 611507
or PBX: (0125) 611377
Telex: (0122) 160 INPE BR

*(3) Canada*

Canada Centre for Remote Sensing (CCRS)
User Assistance and Marketing Unit
717 Belfast Road
Ottawa, Ontario K1A OY7
Canada
Telephone: 613 995-1210
Telex: 053-3777

*(4) Europe*

Earthnet User Services
Via Galileo Galilei
100 44 Frascati, Italy
Telephone: 39-6-9401360 or 39-6-9401216
Telex: 611295 or 610637

*(5) Japan*

Remote Sensing Technology Center of Japan
Uni-Roppongi Bldg, 7-15-17 Roppongi
Minato-ku, Tokyo 106, Japan
Telephone: Tokyo 3-403-1761
Telex: 02426780 RESTECJ

*(6) India*

National Remote Sensing Agency
Balanagar, Hyderabad–500 037
Andhra Pradesh, India
Telephone: 262572 Ext. 67
Telex: 0155-522

*(7) Australia*

Australian Landsat Station
14–16 Oatley Court
PO Box 28,
Belconnen, ACT 2616,
Australia
Telephone: 062-515411
Telex: 61510

*(8) Argentina*

Comision Nacional de Investigaciones Espaciales (CNIE)
Centro de Procesamiento
Dorrego 4010
(1425) Buenos Aires, Argentina
Telephone: 772 5108
Telex: 17511 LANBAR AR

*(9) South Africa*

National Institute for Telecommunications Research
ATTN: Satellite Remote Sensing Centre
PO Box 3718
Johannesburg 2000, South Africa
Telephone: 27-12-26-5271
Telex: 3-21005 SOUTH AFRICA

*(10) Thailand*

Remote Sensing Division
National Research Council
196 Phahonyothin Road
Bangkok 10900, Thailand
Telex: 82213 NRCTRSD
Telephone: 579-0117
Cable: NRC Bangkok

*(11) China*

Academica Sinica
Landsat Ground Station
Beijing, People's Republic of China
Telex: 210222 ASCHI CN
Telephone: 284861 (Beijing, China)

*(12) Indonesia*

Chairman
Indonesian National Institute of Aeronautics and Space
JLN Pemuda Persil NO. 1
PO Box 3048
Djakarta, Indonesia
Telex: 49175

*Source: NOAA 1986*

# Bibliography

ACKLESON, S.G., KLEMAS, V., MCKIM, H.L. and MERRY, C.J. (1985), 'A comparison of SPOT simulator data with Landsat MSS imagery for delineating water masses in Delaware Bay, Broadkill River, and adjacent wetlands', *Photogrammetric Engineering and Remote Sensing* 51, 1123–9.

AL ABBAS, A.H., SWAIN, P.H. and BAUMGARDNER, M.F. (1972), 'Relating organic matter and clay content to the multispectral radiance of soils', *Soil Science* 114, 477–85.

ALLAN, J.A. (1977), 'Land use and changes in land use in the Urla region of Aegean Turkey', in van Genderen, J.L. and Collins, W.G. (eds) *Monitoring Environmental Change by Remote Sensing*, Remote Sensing Society, Reading, 7–14.

ALLAN, J.A. and RICHARDS, T.S. (1983), 'Remote sensing for identifying low density vegetation cover in semi-arid coastal north-west Egypt', *Remote Sensing for Rangeland Monitoring and Management*, Remote Sensing Society, Reading, 69–79.

ALLAN, T.D. (ed.) (1983), *Satellite Microwave Remote Sensing*, Ellis Horwood Ltd, Chichester.

ALLISON, L.J. (1977), *Geological applications of Nimbus radiation data in the Middle East*, NASA Technical Note D-8469, Washington DC.

ALLISON, L.J. and SCHNAPF, A. (1983), 'Meteorological satellites', in Colwell, R.N. (ed.), *Manual of Remote Sensing*, 2nd edn, vol. I, American Society of Photogrammetry, Virginia, 651–79.

ALWASH, M., ZAKIN, F. and BAKOR, A.R. (1986), *Urban Mapping of the City of Jeddah, Kingdom of Saudi Arabia on the Basis of High Resolution Satellite Data*, Faculty of Earth Sciences, King Abdulaziz University.

ANDERSON, J.E. (1985), 'The use of Landsat-4 MSS digital data sets and the evaluation of scene-to-scene registration accuracy', *Photogrammetric Engineering and Remote Sensing* 51, 457–62.

ANDERSON, J.R., HARDY, E.E., ROACH, J.T. and WITMER, R.E. (1976), *A Land Use and Land Cover Classification for Use with Remote Sensor Data*, US Geological Survey Professional Paper 964, Washington, DC.

ANON (1985), 'The Shuttle looks at Earth', *Sky & Telescope* 70, 12–14.

ARANUVACHAPUN, S. (1983), 'Variation of atmospheric optical depth for remote sensing radiance calculations', *Remote Sensing of Environment* 13, 131–47.

ARANUVACHAPUN, S. (1985), 'Satellite remote sensing of ocean colours', *Advanced*

*Technology for Monitoring and Processing Global Environmental Data*, Remote Sensing Society, Reading, 513–21.

ASANUMA, I., MUNEYAMA, K., SASAKI, Y., IISAKA, J., YASUDA, Y. and EMORI, Y. (1986), 'Satellite thermal observations of oil slicks on the Persian Gulf', *Remote Sensing of Environment* 19, 171–86.

ATLAS, D. and THIELE, D.W. (eds) (1981), *Precipitation Measurements from Space. Workshop Report*, NASA Goddard Space Flight Center, Maryland.

BADHWAR, G.D., MACDONALD, R.B. and MEHTA, N.C. (1986), 'Satellite-derived leaf-area-index and vegetation maps as input to global carbon cycle models – a hierarchical approach', *International Journal of Remote Sensing* 7, 265–81.

BAKER, M.C.W. and BALDWIN, J.A. (1981), 'Application of Landsat multispectral classification for locating gossans in north Chile', in Allan, J.A. and Bradshaw, M. (eds), *Geological and Terrain Analysis Studies by Remote Sensing*, Remote Sensing Society, Reading, 25–33.

BARBER, R.C. (1985), 'Theory of digital imaging from orbital synthetic-aperture radar', *International Journal of Remote Sensing* 6, 1009–57.

BARRETT, E.C. (1970), 'The estimation of monthly rainfall from satellite data', *Monthly Weather Review* 101, 215–22.

BARRETT, E.C. (1971), 'The tropical Far East: ESSA satellite evaluations of high season climatic patterns', *Geographical Journal* 137, 535–55.

BARRETT, E.C. (1974), *Climatology from Satellites*, Methuen, London.

BARRETT, E.C. and CURTIS, L.F. (1982), *Introduction to Environmental Remote Sensing*, 2nd edn, Chapman and Hall, London.

BARRETT, E.C. and MARTIN, D.W. (1981), *The Use of Satellite Data in Rainfall Monitoring*, Academic Press, London.

BARRY, R.G. and CHORLEY, R.J. (1982), *Atmosphere, Weather and Climate*, 4th edn, Methuen, London.

BATLIVALA, P.P. and ULABY, F.T. (1977), 'Estimation of soil moisture with radar remote sensing', *Proceedings 11th International Symposium on Remote Sensing of Environment*, Ann Arbor, Michigan, 1557–66.

BENGTSSON, L. (1982), 'Observational requirements for long range forecasting', *WMO Long-range Forecasting Research Publication* 1, World Meteorological Organisation, Geneva, 219–29.

BENNY, A.H. (1983), 'Automatic relocation of ground control points in Landsat imagery', *International Journal of Remote Sensing* 4, 335–42.

BENNY, A.H. (1985), 'An example of the use of Landsat satellite imagery for the accurate location of offshore islands', *International Journal of Remote Sensing* 6, 1581–4.

BERNSTEIN, R. and FERNEYHOUGH, D.G. (1975), 'Digital image processing', *Photogrammetric Engineering* 41, 1465–76.

BORENGASSER, M.X. and TARANIK, J.V. (1985), 'Evaluation of SPOT simulator data for the detection of alteration in Goldfield/Cuprite, Nevada', *Photogrammetric Engineering and Remote Sensing* 51, 1109–14.

BRADBURY, P.A., HAINES-YOUNG, R.H., MATHER, P.M. and MACDONALD, A. (1985), 'The use of remotely-sensed data for landscape classification in Wales: the status of woodlands in the landscape', *Advanced Technology for Monitoring and Processing Global Environmental Data*, Remote Sensing Society, Reading, 401–11.

BROWNING, K.A. (1985), 'Conceptual models of precipitation systems', *ESA Journal* 9, 157–80.
BROWNING, K.A. and COLLIER, C.G. (1982), 'An integrated radar-satellite nowcasting system in the UK', in Browning, K.A. (ed.), *Nowcasting*, Academic Press, London, 47–61.
BRYAN, H.L. (1983), 'Urban land use classification using synthetic aperture radar', *International Journal of Remote Sensing* 4, 215–33.
BRYANT, N.A., ZOBRIST, A.L., WALKER, R.E. and GOKHMANN, B. (1985), 'Landsat Thematic Mapper geodetic accuracy: implications for geocoded map compatibility', *Advanced Technology for Monitoring and Processing Global Environmental Data*, Remote Sensing Society, Reading, 257–73.
BUCHAN, G.M. and HUBBARD, N.K. (1986), 'Remote sensing in land-use planning: an application in west central Scotland using SPOT-simulation data', *International Journal of Remote Sensing* 7, 767–77.
BULLARD, R.K. and DIXON-GOUGH, R.W. (1985), *The Landsat Atlas of Britain*, Taylor & Francis, London.
BURKE, H.H.K., BOWLEY, C.J. and BARNES, J.C. (1984), 'Determination of snowpack properties from satellite passive microwave measurements', *Remote Sensing of Environment* 15, 1–20.

CADET, D. and DESBOIS, M. (1979), 'Low level airflow over the western Indian Ocean as seen from Meteosat', *Nature* 278, 538–9.
CALLISON, R.D. and CRACKNELL, A.P. (1984), 'Atmospheric correction to AVHRR brightness temperatures around Great Britain', *International Journal of Remote Sensing* 5, 185–98.
CANAS, A.A.D. and BARNETT, M.E. (1985), 'The generation and interpretation of false-colour composite principal component images', *International Journal of Remote Sensing* 6, 867–81.
CARAUX, D. and AUSTIN, R.W. (1983), 'Delineation of seasonal changes of chlorophyll frontal boundaries in Mediterranean coastal waters with Nimbus-7 Coastal Zone Color Scanner data', *Remote Sensing of Environment* 13, 239–49.
CARPENTER, D.J. and CARPENTER, S.M. (1983), 'Modeling inland water quality using Landsat data', *Remote Sensing of Environment* 13, 345–52.
CARTER, P. (1979), 'Urban monitoring with LANDSAT imagery', *Machine-Aided Image Analysis 1978*, Institute of Physics, Conference series, number 44, Bristol, 220–30.
CARTER, P. and STOW, B. (1979), 'Clean-up of digital thematic maps of urban growth extracted from Landsat imagery', in Allan, J.A. and Harris, R. (eds), *Remote Sensing and National Mapping*, Remote Sensing Society, Reading, 27–40.
CASSINIS, R., TOSI, N., LECHI, G.M., BRIVIO, P.A., ZILIOLI, E. and MARINI, A. (1984), 'Thermal inertia of rocks – an HCMM experiment on Sardinia, Italy', *International Journal of Remote Sensing* 5, 79–94.
CHESHIRE, H.M., KHORRAM, S. and BROCKHAUS, J.A. (1985), 'Monitoring estuarine water quality from Landsat TM', *Advanced Technology for Monitoring and Processing Global Environmental Data*, Remote Sensing Society, Reading, 227–36.
COLWELL, R.N. (ed.) (1983), *Manual of Remote Sensing*, 2nd edn, 2 vols, American Society of Photogrammetry, Virginia.
CRACKNELL, A.P. (1983), 'Applications for fisheries', *Remote Sensing. New Systems and Potential Applications*, ESA SP-205, European Space Agency, Paris, 101–14.

CURRAN, P.J. (1985), *Principles of Remote Sensing*, Longman, London.
CURRAN, P.J. and WILLIAMSON, H.D. (1985), 'The accuracy of ground data used in remote sensing investigations', *International Journal of Remote Sensing* 6, 1637–51.

DAVIES, P.A. and CHARLTON, J.A. (1986), 'Remote sensing of coastal discharge sites using SPOT-simulation data', *International Journal of Remote Sensing* 7, 815–24.
DAVISON, G.J. (1986), 'Ground control pointing and geometric transformation of satellite imagery', *International Journal of Remote Sensing* 7, 65–74.
DE GLORIA, S.D. (1985), 'Evaluation of simulated SPOT imagery for the interpretation of agricultural resources in California', *Photogrammetric Engineering and Remote Sensing* 51, 1103–8.
DOWMAN, I.J. (1981), 'Topographic mapping from space imagery. Progress and potential', *Matching Remote Sensing Technologies and their Applications*, Remote Sensing Society, Reading, 529–39.
DOYLE, F.J. (1985), 'Large Format Camera photograph of Massachussetts', *Photogrammetric Engineering and Remote Sensing* 51, 200.
DUGGIN, M.J. and PHILIPSON, W.R. (1985), 'Relating ground, aircraft and satellite radiance measurements', *International Journal of Remote Sensing* 6, 1665–70.
DUGGIN, M.J. and SAUNDERS, R.W. (1984), 'Problems encountered in remote sensing of land and ocean surface features', in Henderson-Sellers, A. (ed.), *Satellite Sensing of a Cloudy Atmosphere: Observing the Third Planet*, Taylor & Francis, London, 241–87.
DUGGIN, M.J., SAKHAVAT, H. and LINDSAY, J. (1985), 'The systematic and random variation of recorded radiation in a Landsat Thematic Mapper image', *International Journal of Remote Sensing* 6, 1257–61.

ELACHI, C. (1985), 'Microwave and infrared satellite remote sensors', in Colwell, R.N. (ed.), *Manual of Remote Sensing*, American Society of Photogrammetry, Virginia, 571–650.
ELACHI, C., CIMINO, J.B. and SETTLE, M. (1986), 'Overview of the Shuttle Imaging Radar-B preliminary scientific results', *Science* 232, 1511–16.
ERICKSON, J.D. (1984), 'The LACIE experiment in satellite aided monitoring of global crop production', in Woodwell, G.M. (ed.), *The Role of Terrestrial Vegetation in the Global Carbon Cycle: Measurement by Remote Sensing*, John Wiley and Sons Ltd, Chichester, 191–217.
ESA (1984), *Earth Observation Quarterly* 8, European Space Agency, ESTEC, Noordwijk.
ESA (1985), *Looking Down Looking Forward*, European Space Agency, ESTEC, Noordwijk.
ESA (1986), *Earth Observation Requirements for the Polar Orbiting Platform Elements of the International Space Station*, European Space Agency, Paris.
ESSERY, C.I. and WILCOCK, D.N. (1986), 'SPOT-simulation campaign: a preliminary land-use classification for a 200km$^2$ river catchment', *International Journal of Remote Sensing* 7, 801–814.
EZRA, C., TINNEY, L.R. and JACKSON, R.D. (1984), 'Effect of soil background on vegetation discrimination using Landsat data', *Remote Sensing of Environment* 16, 233–42.

FOLLANSBEE, W.A. (1973), *Estimation of average daily rainfall from satellite cloud photographs*, NOAA Technical Memorandum NESS 44, Washington, DC.

FORD, J.P., CIMINO, J.B. and ELACHI, C. (1982), *Space Shuttle Columbia Views the World with Imaging Radar: the SIR-A Experiment*, JPL Publication 82–95, Jet Propulsion Laboratory, California.

FORSHAW, M.R.B., HASKELL, A., MILLER, P.F., STANLEY, D.J. and TOWNSHEND, J.R.G. (1983), 'Spatial resolving power of remotely sensed imagery. A review paper', *International Journal of Remote Sensing* 4, 497–520.

FORSTER, B.C. (1985), 'An examination of some problems and solutions in monitoring urban areas from satellite platforms', *International Journal of Remote Sensing* 6, 139–51.

FOSTER, J.L., RANGO, A., HALL, D.K., CHANG, A.T.C., ALLISON, L.J. and DIESEN, B.C. (1980), 'Snowpack monitoring in North America and Eurasia using passive microwave satellite data', *Remote Sensing of Environment* 10, 285–98.

FOTHERINGHAM, R.R. (1979), *The Earth's Atmosphere Viewed from Space*, University of Dundee, Dundee.

FREDEN, S.C. and GORDON, F. JR (1983), 'Landsat satellites', in Colwell, R.N. (ed.), *Manual of Remote Sensing*, 2nd edn, vol. I, American Society of Photogrammetry, Virginia, 517–70.

FURNISS, T. (1986), 'Space station. NASA's greatest challenge', *Flight International* 30 August 1986, 137–40.

FUSCO, L., FREI, U., TREVERSE, D., BLONDA, P.N., PASQUARIELLO, G. and MILILLO, G. (1986), 'Landsat TM image forward/reverse scan banding: characterization and correction', *International Journal of Remote Sensing* 7, 557–75.

GALLI DE PARATESI, S. and REINIGER, P. (eds) (1983), *Heat Capacity Mapping Mission. Investigation No. 25 (Tellus Project)*, Final Report, Joint Research Centre, Ispra, Italy.

GELSEMA, E.S. and KANAL, L.N. (eds) (1980), *Pattern Recognition in Practice*, North-Holland, Amsterdam.

GONZALEZ, R.W. (1982), 'Remote sensing techniques used in hydrocarbon exploration. A state-of-the-art review', *Remote Sensing in Exploration Geology*, Proceedings Second Thematic Conference, Ann Arbor, Michigan, vol. II, 615–24.

GORDON, H.R., CLARK, D.L., MUELLER, J.L. and HOVIS, W.A. (1980), 'Phytoplankton pigments from the Nimbus-7 Coastal Zone Color Scanner: comparisons with surface measurements', *Science* 210, 63–6.

GREGORY, K.J. (1985), *The Nature of Physical Geography*, Edward Arnold, London.

GRIFFITH, C.G., WOODLEY, W.L., GRUMBE, P.G., MARTIN, D.W., STOUT, J. and SIKDAR, D.N. (1978), 'Rain estimation from geosynchronous imagery – visible and infrared studies', *Monthly Weather Review* 106, 1153–71.

GRUBER, A. and KRUEGER, A.F. (1984), 'The status of the NOAA outgoing long wavelength radiation data set', *Bulletin American Meteorological Society* 65, 958–62.

HALL, D.K. and MARTINEC, J. (1985), *Remote Sensing of Ice and Snow*, Chapman and Hall, London.

HARALICK, R.M. (1979), 'Statistical and structural approaches to texture', *Proceedings IEEE* 67, 786–804.

HARALICK, R.M., SHANMUGAN, K. and DINSTEIN, I. (1973), 'Textural features for image classification', *IEEE Transactions on Systems, Man and Cybernetics* SMC-3, 610–21.

HARALICK, R.M. and SHANMUGAN, K.S. (1974), 'Combined spectral and spatial processing of ERTS imagery data', *Remote Sensing of Environment* 3, 3–13.

HARDY, J.R. (1985), 'Geometric quality of a Thematic Mapper image of the United Kingdom', *Remote Sensing: Data Acquisition, Management and Applications*, Remote Sensing Society, Reading, 3–12.

HARDY, J.R. and AGAR, C.D. (1978), 'A study of the potential of Landsat MSS digital data for woodland census in Britain', *Journal of the British Interplanetary Society* 31, 467–74.

HARRIES, J.E., LLEWELLYN-JONES, D.T., MINNETT, P.J., SAUNDERS, R.W. and ZAVODY, A.M. (1983), 'Observations of sea-surface temperature for climate research', *Philosophical Transactions Royal Society of London* A, 309, 381–95.

HARRIS, R. (1980), 'Spectral and spatial image processing for remote sensing', *International Journal of Remote Sensing* 1, 361–75.

HARRIS, R. (1983), 'Remote sensing support for the Omani White Oryx Project', *Remote Sensing for Rangeland Monitoring and Management*, Remote Sensing Society, Reading, 17–24.

HARRIS, R. (1985a), 'Satellite remote sensing: low spatial resolution', *Progress in Physical Geography* 9, 600–6.

HARRIS, R. (1985b), 'SIR-A imagery of Tunisia and its potential for population estimation', *International Journal of Remote Sensing* 6, 975–8.

HARRIS, R. and BARRETT, E.C. (1978), 'Toward an objective nephanalysis', *Journal of Applied Meteorology* 17, 1258–66.

HASKELL, A. (1983), 'The ERS-1 programme of the European Space Agency', *ESA Journal* 7, 1–13.

HAYES, L. (1985), 'The current use of TIROS-N series of meteorological satellites for land-cover studies', *International Journal of Remote Sensing* 6, 35–45.

HAYES, L. and CRACKNELL, A.P. (1984), 'Vegetation depiction by AVHRR – a Scottish sampling', *Satellite Remote Sensing – Review and Preview*, Remote Sensing Society, Reading, 181–90.

HEILMAN, J.L. and MOORE, D.G. (1982), 'Evaluating near-surface soil moisture using Heat Capacity Mapping Mission (HCMM) data', *Remote Sensing of Environment* 12, 117–21.

HENDERSON-SELLERS, A. (ed.) (1984), *Satellite Sensing of a Cloudy Atmosphere: Observing the Third Planet*, Taylor & Francis, London.

HILDEBRANDT, G. (1983), 'Remote sensing from space for forestry purposes', *Remote Sensing. New Satellite Systems and Potential Applications*, ESA SP-205, ESA, Paris, 67–74.

HOEKMAN, D.H. (1985), 'Radar backscattering of forest stands', *International Journal of Remote Sensing* 6, 325–43.

HOFFER, R.M. (1978), 'Biological and physical considerations in applying computer-aided analysis techniques to remote sensor data', in Swain, P.H. and Davis, S.M. (eds), *Remote Sensing. The Quantitative Approach*, McGraw-Hill, New York, 227–89.

HORLER, D.N.H. and AHERN, F.J. (1986), 'Forestry information content of Thematic Mapper data', *International Journal of Remote Sensing* 7, 405–28.

HORNE, A.I.D. (1984), 'Forest cover monitoring by remote sensing in Great Britain', *Integrated Approaches in Remote Sensing*, European Association of Remote Sensing Laboratories, Paris, 99–107.

HOUGHTON, J.T. (1985), 'Satellite meteorology', *Recent Advances in Meteorology and Physical Oceanography*, Royal Meteorological Society, Bracknell, 47–58.

HOUGHTON, J.T., TAYLOR, F.W., and RODGERS, C.D. (1984), *Remote Sounding of Atmospheres*, Cambridge University Press, Cambridge.

HUBBARD, N. and WRIGHT, R. (1982), 'A semi-automatic approach to land cover classification of Scotland from Landsat', *Remote Sensing and the Atmosphere*, Remote Sensing Society, Reading, 212–21.

HUETE, A.R., POST, D.F. and JACKSON, R.D. (1984), 'Soil spectral effects on 4-space vegetation discrimination', *Remote Sensing of Environment* 15, 155–65.

HUNT, G.R. and SALISBURY, J.W. (1970), 'Visible and near-infrared spectra of minerals and rocks: I Silicate minerals', *Modern Geology* 1, 283–300.

HUNT, G.R. and SALISBURY, J.W. (1976), 'Visible and near-infrared spectra of minerals and rocks: XI Sedimentary rocks', *Modern Geology* 5, 211–17.

HUNT, G.R., SALISBURY, J.W. and LENHOFF, C.J. (1973), 'Visible and near-infrared spectra of minerals and rocks: VII Acidic igneous rocks', *Modern Geology* 4, 85–106.

IRANPANAH, I. (1977), 'Geologic applications of Landsat imagery', *Photogrammetric Engineering and Remote Sensing* 43, 1037–40.

IRONS, J.R., MARKHAM, B.L., NELSON, R.F., TOLL, D.L., WILLIAMS, D.L., LATTY, R.S. and STAUFFER, M.L. (1985), 'The effects of spatial resolution on the classification of Thematic Mapper data', *International Journal of Remote Sensing* 6, 1385–1403.

JACKSON, M.J. and MASON, D.C. (1986), 'The development of integrated geoinformation systems', *International Journal of Remote Sensing* 7, 723–40.

JACOBOWITZ, H. and TIGHE, R.J. (1984), 'The ERB derived from the Nimbus-7 ERB experiment', *Journal of Geophysical Research* 89, 4997–5010.

JAQUES, B. and LOPEZ, R. (1986), 'Looking down for money', *Space Markets* 2, 73–80.

JENSEN, J.R. (1983), 'Urban/suburban land use analysis', in Colwell, R.N. (ed.), *Manual of Remote Sensing*, 2nd edn, vol. II, American Society of Photogrammetry, Virginia, 1571–666.

JERRETT, D., EYRE, J.R. and MCCALLUM, E. (1982), 'High resolution soundings of temperature in the European/North Atlantic area', *Remote Sensing and the Atmosphere*, Remote Sensing Society, Reading, 230–7.

JOHNSON, G.L. and SUCHMAN, D. (1980), 'Intercomparisons of SMS wind sets: a study using rapid-scan imagery', *Monthly Weather Review* 108, 1672–88.

JOHNSTON, R.J. (1980), *Multivariate Statistical Analysis in Geography*, Longman, London.

JUSTICE, C.O. and TOWNSHEND, J.R.G. (1981), 'Integrating ground data with remote sensing', in Townshend, J.R.G. (ed.), *Terrain Analysis and Remote Sensing*, George Allen & Unwin, London, 38–58.

JUSTICE, C.O., TOWNSHEND, J.R.G., HOLBEN, B.N. and TUCKER, C.J. (1985), 'Analysis of the phenology of global vegetation using meteorological satellite data', *International Journal of Remote Sensing* 6, 1271–318.

KANAL, L.N. and ROSENFELD, A. (eds) (1981), *Progress in Pattern Recognition*, vol. 1, North-Holland, Amsterdam.

KÄSTNER, M., FISCHER, H. and BOLLE, H.J. (1980), 'Wind determination from Nimbus 5 observations in the 6.3 μm water vapor band', *Journal of Applied Meteorology* 19, 409–18.

KAUTH, R.G. and THOMAS, G.S. (1976), 'The tasseled cap, a graphic description of the spectral–temporal development of agricultural crops as seen by Landsat',

*Proceedings of the Symposium on Machine Processing of Remotely Sensed Data*, Purdue, Indiana, 6.23–7.20.

KHORRAM, S. (1981), 'Water quality mapping from Landsat digital data', *International Journal of Remote Sensing* 2, 145–53.

KHORRAM, S. (1982), 'Remote sensing of salinity in the San Francisco Bay delta', *Remote Sensing of Environment* 12, 15–22.

KHORRAM, S. and CHESHIRE, H.M. (1985), 'Remote sensing of water quality in the Neuse river estuary, North Carolina', *Photogrammetric Engineering and Remote Sensing* 51, 329–41.

KIRSCHNER, F.G., KAMINSKY, S.A., WEISMILLER, R.A., SINCLAIR, H.R. and HINZEL, E.J. (1978), 'Map unit assessment using drainage classes defined by Landsat data', *Soil Science Society of America Journal* 42, 768–71.

KITTLER, J., FU, K.S. and PAU, L.F. (1981), *Pattern Recognition Theory and Applications*, Reidel, Dordrecht.

KLEMAS, V. (1980), 'Remote sensing of coastal fronts and their effects on oil dispersion', *International Journal of Remote Sensing* 1, 11–28.

KLEMAS, V. (1981), 'Technology transfer to developing countries: future use of remote sensing in biological marine resource development', *Workshop on Future of International Cooperation in Marine Technology*, Ocean Policy Committee, National Academy of Sciences, University of Delaware, Newark.

KOBAYASHI, T. and HIROSAWA, H. (1985), 'Measurement of radar backscatter from rough soil surfaces using linear and circular polarizations', *International Journal of Remote Sensing* 6, 345–52.

KRINSLEY, D.B. (1974), *The Utilization of ERTS-1 Generated Images in the Evaluation of Some Iranian Playas as Sites for Economic and Engineering Development*, NASA Final Technical Report, Contract S-70243-AG-3, Washington DC.

KRISTOF, S.J. and ZACHARY, A.L. (1974), 'Mapping soil features from MSS data', *Photogrammetric Engineering* 40, 1427–34.

LABOVITZ, M.L., MASUOKA, E.J., BELL, R., NELSON, R.F., LARSEN, C.A., HOOKER, L.K. and TROENSEGAARD, K.W. (1985), 'Experimental evidence for spring and autumn windows for the detection of geobotanical anomalies through the remote sensing of overlying vegetation', *International Journal of Remote Sensing* 6, 195–216.

LAKE, S.D., MUNDAY, T.J. and DEWEY, J.F. (1984), 'Lineament mapping and analysis in the Wessex basin of southern England: a comparison between MSS and TM data', *Satellite Remote Sensing. Review and Preview*, Remote Sensing Society, Reading, 361–74.

LASKER, R., PELÁEZ, J. and LAURS, R.M. (1981), 'The use of satellite infrared imagery for describing ocean processes in relation to spawning of the Northern Anchovy (*Engraulis mordax*)', *Remote Sensing of Environment* 11, 439–53.

LATHAM, J.S. and HUBBARD, N.K. (1985), 'Landsat-5 Thematic Mapper scene of Abu Dhabi', *International Journal of Remote Sensing* 6, 1491–2.

LAWTON, R.M. (1978), *Vegetation reconnaissance survey for the Oryx Project. Report to the Sultanate of Oman and the World Wildlife Fund*, Ministry of Overseas Development, Surrey, Project report 55.

LILJAS, E. (1982), 'Automated techniques for the analysis of satellite cloud imagery', in Browning, K.A. (ed.), *Nowcasting*, Academic Press, London, 167–76.

LINDELL, L.T., STEINVALL, O., JONSSON, M. and CLAESSON, TH. (1985), 'Mapping of coastal-water turbidity using LANDSAT imagery', *International Journal of Remote Sensing* 6, 629–42.

LINTZ, J. and SIMONETT, D.S. (eds) (1976), *Remote Sensing of Environment*, Addison-Wesley, Reading, Massachusetts.

LODGE, D.W.S. (1983), 'Expressions of bathymetry on SEASAT synthetic aperture radar images', in Allan, T.D. (ed.), *Satellite Microwave Remote Sensing*, Ellis Horwood Ltd., Chichester, 247–60.

LODGE, D.W.S. (1985), 'Using space stations: not quite so remote sensing', *Advanced Technology for Monitoring and Processing Global Environmental Information*, Remote Sensing Society, Reading, 29–38.

LÖFFLER, E. and MARGULES, C. (1980), 'Wombats detected from space', *Remote Sensing of Environment* 9, 47–56.

LOGAN, T.L. (1983), 'Regional forest biomass modelling from combined AVHRR and Landsat data', *1983 National Conference on Resource Management Applications: Energy and Environment*, CERMA, Virginia, vol. III, 49–64.

LULLA, K. (1985), 'Some observations on geobotanical remote sensing and mineral prospecting', *Canadian Journal of Remote Sensing* 11, 17–38.

MAMULA, N. and VOIGHT, B. (1982), 'Structural analysis of a divergent plate boundary terrain – a Landsat perspective', *Remote Sensing for Exploration Geology*, Proceedings Second Thematic Conference, Ann Arbor, Michigan, vol. I, 385–416.

MARKHAM, B.L. and BARKER, J.L. (1985a), 'Foreward. Special Issue LIDQA Final Symposium', *Photogrammetric Engineering and Remote Sensing* 51, 1245–6.

MARKHAM, B.L. and BARKER, J.L. (1985b), 'Spectral characterization of Landsat Thematic Mapper sensors', *International Journal of Remote Sensing* 6, 697–716.

MASSIN, J.M. (ed.) (1984), *Remote Sensing for the Control of Marine Pollution*, Plenum, New York.

MATTIE, M.G., LICHY, D.E. and BEAL, R.C. (1980), 'Seasat detection of waves, currents and inlet discharges', *International Journal of Remote Sensing* 1, 377–98.

MAUL, G.A. (1985), *Introduction to Satellite Oceanography*, Martinus Nijhoff Publishers, Dordrecht.

MILTON, E.J. (1980), 'A portable multiband radiometer for ground data collection in remote sensing', *International Journal of Remote Sensing* 1, 153–65.

MILTON, E.J. and WEBB, J.P. (1984), 'The use of narrow bandwidth spectro-radiometry for soils investigations', *Satellite Remote Sensing. Review and Preview*, Remote Sensing Society, Reading, 401–10.

MITCHELL, C.W. (1981), 'Reconnaissance land resource surveys in arid and semi-arid environments', in Townshend, J.R.G. (ed.), *Terrain Analysis and Remote Sensing*, George Allen and Unwin, London, 169–83.

MOORE, J.MCM. and CAMM, S. (1982), 'Interactive enhancement of Landsat imagery for structural mapping in tin-tungsten prospecting: a case history of the SW England orefield (UK)', *Remote Sensing for Exploration Geology*, Proceedings Second Thematic Conference, Ann Arbor, Michigan, vol. II, 727–40.

MOORE, R.K. (1983), 'Imaging radar systems', in Colwell, R.N. (ed.), *Manual of Remote Sensing*, 2nd edn, vol. I, American Society of Photogrammetry, Virginia, 429–74.

MORRISELY, L.A., WEINSTOCK, K.J., MOUAT, D.A. and CARD, D.H. (1984), 'Statistical analysis of Thematic Mapper simulator data for the geobotanical discrimination of rock types in south west Oregon', *IEEE Transactions on Geoscience and Remote Sensing* GE-22, 525–9.

MUIRHEAD, K. and CRACKNELL, A.P. (1984), 'Identification of gas flares in the North Sea using satellite data', *International Journal of Remote Sensing* 5, 199–212.

MUIRHEAD, K. and CRACKNELL, A.P. (1985), 'Straw burning over Great Britain detected by AVHRR', *International Journal of Remote Sensing* 6, 827–33.

MUNDAY, J.C. JR. and ALFÖLDI, T.T. (1979), 'Landsat test of diffuse reflectance models for aquatic suspended solids measurements', *Remote Sensing of Environment* 8, 169–83.

MYERS, V.I. (1983), 'Remote sensing applications in agriculture', *Manual of Remote Sensing* (ed. R.N. Colwell), 2nd edition, vol. II, American Society of Photogrammetry, Virginia, 2111–228.

NASA (1978), *HCMM User's Guide*, Goddard Space Flight Center, Maryland.

NASA (1984), *Earth Observing System. Working Group Report Volume I*, NASA, Goddard Space Flight Center, Maryland, NASA TM-86129.

NATIONAL REMOTE SENSING CENTRE (1985), *Data Users Guide*, RAE, Farnborough.

NEEDHAM, B.H. (1984), 'NOAA satellite infrared images of the Gulf Stream', *International Journal of Remote Sensing* 5, 241–2.

NELSON, R. and HOLBEN, B. (1986), 'Identifying deforestation in Brazil using multiresolution satellite data', *International Journal of Remote Sensing* 7, 429–48.

NERC (1985), *The Natural Environment Research Council. Report of the Council for the period 1 April 1984–31 March 1985*, NERC, Swindon.

NOAA (1986), *Landsat Data Users Notes*, Issue No. 35, South Dakota, USA.

NRSC (1985), *ERS-1*, National Remote Sensing Centre, Farnborough.

OHRING, G. and SELLERS, P. (1984), *Report of the Design Workshop for the First ISLSCP Field Experiment*, ISLSCP Report No. 3, NOAA/NASA, Washington DC.

OLIVER, V.J. and SCOFIELD, R.A. (1976), 'Estimation of rainfall from satellite imagery', *Sixth Conference on Weather Forecasting and Analysis*, American Meteorological Society, Boston, 242–5.

ORTI, F., GARCIA, A. and MARTIN, M.A. (1979), 'Geometric correction of Landsat MSS images using a ground control point library', in Allan, J.A. and Harris, R. (eds), *Remote Sensing and National Mapping*, Remote Sensing Society, Reading, 17–26.

PARRY, D.E. (1978), 'Some examples of the use of satellite imagery (Landsat) for natural resource mapping in western Sudan', in Collins, W.G. and van Genderen, J.L. (eds), *Remote Sensing Applications in Developing Countries*, Remote Sensing Society, Reading, 1–12.

PESLEN, C.A. (1980), 'Short-interval SMS wind vector determinations for a severe local storms area', *Monthly Weather Review* 108, 1407–18.

PODWYSOCKI, M.H., POWER, M.S. and JONES, O.D. (1985), 'Preliminary evaluation of the Landsat-4 Thematic Mapper data for mineral exploration', in Ungar, S.G. (ed.), *The Earth's Surface Studied from Space*, Advances in Space Research, COSPAR, Pergamon Press, Oxford, 13–20.

RANEY, R.K. (1983), 'Synthetic aperture radar observations of ocean and land', *Philosophical Transactions Royal Society London* A 309, 315–21.

RICHARDSON, A.J. (1984), 'El Chichon volcanic ash effects on atmospheric haze measured by NOAA 7 AVHRR data', *Remote Sensing of Environment* 16, 157–64.

ROBIN, G. DE Q., DREWRY, D.J. and SQUIRE, V.A. (1983), 'Satellite observations of polar ice fields', *Philosophical Transactions Royal Society of London* A 309, 447–61.
ROBINSON, I.S. (1983), 'Satellite observations of ocean colour', *Philosophical Transactions Royal Society of London* A 309, 415–32.
ROBINSON, I.S., WELLS, N.C. and CHARNOCK, H. (1984), 'The sea surface thermal boundary layer and its relevance to the measurement of sea surface temperature by airborne and spaceborne radiometers', *International Journal of Remote Sensing* 5, 19–45.
ROTHERY, D.A. and MILTON, E.J. (1981), 'Lithological discrimination in an ophiolite terrain: Landsat MSS imagery and reflectance measurements in Oman', in Allan, J.A., and Bradshaw, M. (eds), *Geological and Terrain Analysis Studies by Remote Sensing*, Remote Sensing Society, Reading, 3–23.

SABINS, F.F. JR. (1978), *Remote Sensing. Principles and Interpretation*, W.H. Freeman and Co., San Francisco.
SCHANDA, E., MÄTZLER, C. and KÜNZI, K. (1983), 'Microwave remote sensing of snow cover', *International Journal of Remote Sensing* 4, 149–58.
SCHLATTER, T.W. (1985), 'A day in the life of a modern mesoscale forecaster', *ESA Journal* 9, 235–56.
SCHMETZ, J. and TURPEINEN, O. (1986), 'Meteorological monitoring with Meteosat', *Abstracts ESA/EARSeL Symposium on Europe from Space*, ESA SP-258, ESA, Paris.
SCHNEIDER, S.R., MCGINNIS, D.F. JR. and STEPHENS, G. (1985), 'Monitoring Africa's Lake Chad with LANDSAT and NOAA satellite data', *International Journal of Remote Sensing* 6, 59–73.
SCOFIELD, R.A. and OLIVER, V.J. (1977), *A scheme for estimating convective rainfall from satellite imagery*, NOAA Technical Memorandum NESS 86, Washington DC.
SELLERS, P.J. (1985), 'Canopy reflectance, photosynthesis and transpiration', *International Journal of Remote Sensing* 6, 1335–72.
SHEFFIELD, C. (1981), *Earthwatch*, Sidgwick & Jackson, London.
SHEFFIELD, C. (1983), *Man on Earth*, Sidgwick & Jackson, London.
SHORT, N. (1982), *The Landsat Tutorial Workbook*, National Aeronautics and Space Administration, Washington DC, NASA Reference Publication 1078.
SILVA, L.F. (1978), 'Radiation and instrumentation in remote sensing', *Remote Sensing. The Quantitative Approach*, McGraw-Hill, New York, 21–135.
SINGH, A. (1984), 'Some clarifications about the pairwise divergence measure in remote sensing', *International Journal of Remote Sensing* 5, 623–7.
SMITH, J.A. (1983), 'Matter–energy interaction in the optical region', in Colwell, R.N. (ed.), *Manual of Remote Sensing*, 2nd edn, vol. I, American Society of Photogrammetry, Virginia, 61–113.
SMITH, W.L. and KELLY, G.A.M. (1985), 'Use of satellite imagery and soundings in mesoscale analysis and forecasting', *ESA Journal* 9, 115–23.
SMITH, W.L., BISHOP, W.P., DVORAK, V.F., HAYDEN, C.M., MCELROY, J.H., MOSHER, F.R., OLIVER, V.J., PURDOM, J.F. and WARK, D.Q. (1986), 'The meteorological satellite: overview of 25 years of operation', *Science* 231, 455–62.
SPOT IMAGE (1985), *SPOT Newsletter* 7, 15 July 1985, Toulouse.
SPOT IMAGE (1986), *SPOT Newsletter* 8, 15 July 1986, Toulouse.

STEFOULI, M. and OSMASTON, H.A. (1984), 'The remote sensing of geological linear features using Landsat: matching analytical approaches to practical applications', *Satellite Remote Sensing. Review and Preview*, Remote Sensing Society, Reading, 227–36.

STRAHLER, A.H. (1981), 'Stratification of natural vegetation for forest and rangeland inventory using Landsat digital imagery and collateral data', *International Journal of Remote Sensing* 2, 15–41.

SUITS, G.H. (1983), 'The nature of electromagnetic radiation', in Colwell, R.N. (ed.), *Manual of Remote Sensing*, 2nd edn, vol. I, American Society of Photogrammetry, Virginia, 37–60.

SUSSKIND, J. (1984), 'Vertical temperature sounding of the atmosphere', in Henderson-Sellers, A. (ed.), *Satellite Sensing of a Cloudy Atmosphere: Observing the Third Planet*, Taylor & Francis, London, 167–202.

SWAIN, P.H. and DAVIS, S.M. (eds) (1978), *Remote Sensing. The Quantitative Approach*, McGraw-Hill, New York.

TARPLEY, J.D., SCHNEIDER, S.R. and MONEY, R.L. (1984), 'Global vegetation indices from NOAA-7 meteorological satellite', *Journal of Climate and Applied Meteorology* 23, 491.

THOMAS, J.O. and DAVEY, P.G. (eds) (1977), *Texture Analysis*, University Physics Photographic Unit, Oxford.

THOMPSON, D.R., HASS, R.A. and MILFORD, M.H. (1981), 'Evaluation of Landsat MSS data for mapping vegetated soil landscapes', *Soil Science Society of America Journal* 45, 91–5.

TOGLIATTI, G. (1986), 'L.F.C.: the second generation photogrammetric camera for space cartography', *Abstracts ESA/EARSeL Symposium on Europe from Space*, ESA SP-258, ESA, Paris.

TOLL, D.L. (1985), 'Analysis of digital Landsat MSS and Seasat SAR data for use in discriminating land cover at the urban fringe of Denver, Colorado', *International Journal of Remote Sensing* 6, 1209–29.

TOU, J.T. and GONZALEZ, R.C. (1974), *Pattern Recognition Principles*, Addison-Wesley, Reading, Massachusetts.

TOWNSHEND, J.R.G. (1984), 'Agricultural land-cover discrimination using thematic mapper spectral bands', *International Journal of Remote Sensing* 5, 681–98.

TOWNSHEND, J.R.G. and TUCKER, C.J. (1981), 'Utility of AVHRR of NOAA 6 and 7 for vegetation mapping', *Matching Remote Sensing Technologies and their Applications*, Remote Sensing Society, Reading, 97–109.

TOWNSHEND, J.R.G., GAYLER, J.R., HARDY, J.R., JACKSON, M.J. and BAKER, J.R. (1983), 'Preliminary analysis of LANDSAT-4 thematic mapper products', *International Journal of Remote Sensing* 4, 817–28.

TUCKER, C.J. (1979), 'Red and photographic infrared linear combinations for mapping vegetation', *Remote Sensing of Environment* 8, 127–50.

TUCKER, C.J. and MATSON, M. (1985), 'Determination of volcanic dust deposition from El Chichon using ground and satellite data', *International Journal of Remote Sensing* 6, 619–27.

TUCKER, C.J., HOLBEN, B.N. and GOFF, T.E. (1984), 'Intensive forest clearing in Rondonia, Brazil, as detected by satellite remote sensing', *Remote Sensing of Environment* 15, 255–61.

TUCKER, C.J., TOWNSHEND, J.R.G. and GOFF, T.E. (1985), 'African land-cover classification using satellite data', *Science* 227, 369–75.

TUCKER, C.J., FUNG, I.Y., KEELING, C.D. and GAMMON, R.H. (1986), 'Relationship between atmospheric $CO_2$ variations and a satellite-derived vegetation index', *Nature* 319, 195–9.

ULBRICHT, K.A. (1983a), 'Comparative experimental study on the use of original and compressed multispectral Landsat data for applied research', *International Journal of Remote Sensing* 4, 571–82.

ULBRICHT, K.A. (1983b), 'LANDSAT image of blue green algae in the Baltic Sea', *International Journal of Remote Sensing* 4, 801–2.

VASS, P.A. (1983), 'A Landsat study of vegetation and seasonal livestock grazing in the southern Sudan', *Remote Sensing for Rangeland Monitoring and Management*, Remote Sensing Society, Reading, 51–68.

WARDLEY, N.W. and CURRAN, P. (1984), 'The estimation of green-leaf-area index from remotely sensed airborne multispectral scanner data', *International Journal of Remote Sensing* 5, 671–9.

WEBER, C. (1985), 'Geological remote sensing: Quo vadis?', *ITC Journal* 4, 227–41.

WEISNET, D.R. and MATSON, M. (1983), 'Remote sensing of weather and climate', in Colwell, R.N. (ed.), *Manual of Remote Sensing*, 2nd edn, vol. II, American Society of Photogrammetry, Virginia, 1305–69.

WELCH, R. (1985), 'Cartographic potential of SPOT image data', *Photogrammetric Engineering and Remote Sensing* 51, 1085–91.

WELCH, R., JORDAN, T.R. and EHLERS, M. (1985), 'Comparative evaluations of the geodetic accuracy and cartographic potential of Landsat-4 and Landsat-5 Thematic Mapper image data', *Photogrammetric Engineering and Remote Sensing* 51, 1249–62.

WESZKA, J.S., DYER, C.R. and ROSENFELD, A. (1976), 'A comparative study of texture measures for terrain classification', *IEEE Transactions Systems, Man and Cybernetics* SMC-6, 269–85.

WILHEIT, T., CHANG, A.T.C., RAO, M.S.V., RODGERS, E.B. and THEON, J.S. (1977), 'Satellite technique for quantitatively mapping rainfall rates over the oceans', *Journal of Applied Meteorology* 16, 551–60.

WILLIAMS, D.F. (1984), 'Overview of the NERC airborne thematic mapper campaign of September 1982', *International Journal of Remote Sensing* 5, 631–4.

WILLIAMS, R.S. JR. (1983), 'Geological applications', in Colwell, R.N. (ed.), *Manual of Remote Sensing*, 2nd edn, vol. II, American Society of Photogrammetry, Virginia, 1667–953.

WILSON, J. and ROESLI, H.P. (1985), 'Use of Doppler radar and radar networks in mesoscale analysis and forecasting', *ESA Journal* 9, 125–46.

WOODLEY, W.L. and SANCHO, B. (1971), 'A first step toward rainfall estimation from satellite cloud photographs', *Weather* 26, 279–89.

WOODLEY, W.L., GRIFFITH, C.G., GRIFFIN, J.S. and STROMATT, S.C. (1980), 'The inference of GATE convective rainfall from SMS-1 imagery', *Journal of Applied Meteorology* 19, 388–408.

WRIGHT, G.G. and BIRNIE, R.V. (1986), 'Detection of surface soil variation using high resolution satellite data: results from the UK SPOT-simulation investigation', *International Journal of Remote Sensing* 7, 757–66.

WU, L.K., MOORE, R.K., ZOUGHI, R., AFIFI, A. and ULABY, F.T. (1985), 'Preliminary results on the determination of the sources of scattering from vegetation canopies at 10GHz. I. Corn and milo', *International Journal of Remote Sensing* 6, 299–313.

YATES, H., STRONG, A., MCGINNIS, D. JR. and TARPLEY, D. (1986), 'Terrestrial observations from NOAA operational satellites', *Science* 231, 463–70.

YOOL, S.R., STAR, J.L., ESTES, J.E., BOTKIN, D.B., ECKHARDT, D.W. and DAVIS, F.W. (1986), 'Performance analysis of image processing algorithms for classification of natural vegetation in the mountains of Southern California', *International Journal of Remote Sensing* 7, 683–702.

YOST, E. and WENDEROTH, S. (1971), 'The reflectance spectra of mineralized trees', *Proceedings Seventh International Symposium on Remote Sensing of Environment*, Ann Arbor, Michigan, 269–84.

ZHENG, Q.A. and KLEMAS, V. (1982), 'Determination of winter temperature patterns, fronts, and surface currents in the Yellow Sea and East China Sea from satellite imagery', *Remote Sensing of Environment* 12, 201–18.

ZHENG, Q., KLEMAS, V. and HUANG, N.E. (1984), 'Dynamics of the slope water off New England and its influence on the Gulf Stream as inferred from satellite IR data', *Remote Sensing of Environment* 15, 135–53.

ZOUGHI, R., MOORE, R.K., ULABY, F.T., WU, L.K. and AFIFI, A. (1985), 'Determination of backscatter sources in surface type targets', *International Journal of Remote Sensing* 6, 285–97.

# Index

LADY MARGARET HALL LIBRARY
OXFORD